## Aims and Scope

The series "Springer Theses" brings together a selection of the very best Ph.D. theses from around the world and across the physical sciences. Nominated and endorsed by two recognized specialists, each published volume has been selected for its scientific excellence and the high impact of its contents for the pertinent field of research. For greater accessibility to non-specialists, the published versions include an extended introduction, as well as a foreword by the student's supervisor explaining the special relevance of the work for the field. As a whole, the series will provide a valuable resource both for newcomers to the research fields described, and for other scientists seeking detailed background information on special questions. Finally, it provides an accredited documentation of the valuable contributions made by today's younger generation of scientists.

## Theses are accepted into the series by invited nomination only and must fulfill all of the following criteria

- They must be written in good English.
- The topic should fall within the confines of Chemistry, Physics, Earth Sciences, Engineering and related interdisciplinary fields such as Materials, Nanoscience, Chemical Engineering, Complex Systems and Biophysics.
- The work reported in the thesis must represent a significant scientific advance.
- If the thesis includes previously published material, permission to reproduce this must be gained from the respective copyright holder.
- They must have been examined and passed during the 12 months prior to nomination.
- Each thesis should include a foreword by the supervisor outlining the significance of its content.
- The theses should have a clearly defined structure including an introduction accessible to scientists not expert in that particular field.

More information about this series at http://www.springer.com/series/8790

Chris Skipper

# Fast Spectral Variability in the X-ray Emission of Accreting Black Holes

Doctoral Thesis accepted by
the University of Southampton, UK

Springer

*Author*
Dr. Chris Skipper
University of Southampton
Southampton
UK

*Supervisor*
Prof. Ian McHardy
University of Southampton
Southampton
UK

ISSN 2190-5053 ISSN 2190-5061 (electronic)
ISBN 978-3-319-09586-8 ISBN 978-3-319-09587-5 (eBook)
DOI 10.1007/978-3-319-09587-5

Library of Congress Control Number: 2014947645

Springer Cham Heidelberg New York Dordrecht London



Printed on acid-free paper

Springer is part of Springer Science+Business Media (www.springer.com)

*The most beautiful thing we can experience is the mysterious. It is the source of all true art and all science. He to whom this emotion is a stranger, who can no longer pause to wonder and stand rapt in awe, is as good as dead: his eyes are closed.*

Albert Einstein (1879–1955)

*To the memory of my grandparents, Arthur and Hazel Jones.*

# Supervisor's Foreword

X-ray emission from accreting compact objects such as black holes is generally assumed to arise from Compton up-scattering of low energy seed photons by very energetic electrons in some sort of corona around the black hole. However, the origin of the seed photons, and whether that origin changes with accretion rate, is not well understood.

Observations of short timescale X-ray spectral variability are crucial for understanding the origin of the seed photons but characteristic variability timescales are very short, e.g., the light travel time across the likely size of the corona in a black hole X-ray binary system is $\sim$millisecond. However although X-ray flux variations have been measured on millisecond timescales, very few X-ray spectral measurements have been carried out on such timescales.

Skipper addresses the problem of understanding X-ray spectral variability on very short timescales in X-ray binaries in a highly innovative manner. He fits simple models to over 2 million X-ray binary spectra on timescales as short as 16 ms and thereby produces lightcurves of the variation of the X-ray power law spectral index. He then cross-correlates the spectral index with the count rate and produces a number of beautiful and insightful results.

He shows that in 'soft' states, where the accretion rate is high, there is a strong positive correlation between X-ray spectral index and count rate, with no measureable lag, i.e., the spectral index becomes 'softer' or steeper as the count rate rises. This observation is explained if the seed photons come predominantly from the surrounding accretion disc and a rise in seed photon flux cools the corona.

However in the 'hardest' states, where the accretion rate is low, there is a strong anti-correlation between X-ray spectral index and count rate, again with lag very close to zero. Skipper interprets these observations in terms of injection of energy into the coronal electron population. This fresh energy hardens the electron energy spectrum and leads to increased production of cyclo-synchrotron seed photons and hardening of the resultant scattered X-ray spectrum.

In between the soft and hard states Skipper shows a smooth transition in the X-ray spectral index versus count rate cross-correlation function, with very interesting lags, which vary with source hardness. This transition can only be explained if there is an energy gradient in the coronal electron population, with higher energy electrons being preferentially located nearer to the black hole. The transition also requires that the variations themselves are produced by perturbations, perhaps in accretion rate, propagating inwards through the corona. Thus he addresses not just the variable origin of the seed photons, but also the structure of the X-ray emission region and the overall origin of the X-ray variations.

Using the same very fast X-ray spectral fits, Skipper also shows very clearly how the X-ray spectral index, in individual X-ray binary sources, changes smoothly from steep to flat and back to steep as the accretion rate increases. This change is not clearly seen in spectra taken over longer timeperiods but fits nicely within the same physical framework as described above.

Skipper extends his work to the study of active galactic nuclei (AGN) over comparable mass-scaled timescales, finding a close similarity between AGN and X-ray binaries.

As a separate part of his thesis, Skipper examines the multiband nuclear properties of a well-defined sample of nearby optically bright galaxies (the 'Palomar Sample'). He finds a number of interesting results regarding low luminosity AGN. In particular, the 'Unified Model' of AGN, whereby obscuration is orientation-dependent, does not seem to apply.

Overall, this thesis is a very good addition to our understanding of X-ray spectral variability, and the structure of the emission region, in accreting black holes.

Southampton, UK, June 2014 Prof. Ian McHardy

# Preface

The discovery of black holes in space represents one of the great scientific advancements of the twentieth century, and continues to this day to stimulate the imaginations of scientists and science-fiction writers alike. The influence of these extraordinary objects upon their environments—effected through the processes of accretion and, perhaps less intuitively, the ejection of matter back out into space—has come to be understood as fundamental in the formation and evolution of the Universe. Furthermore, the extreme gravitational forces that act within the vicinity of each black hole provide a wonderful opportunity for us to refine our understanding of the laws of nature by examining their action under conditions that would simply be impossible to reproduce in a laboratory here on Earth. Therefore, black holes are, and will continue to be, some of the most important and popular targets for scientific study across a wide range of research disciplines.

This thesis was written during my time studying as a Ph.D. student at the Astronomy Group of the University of Southampton, and utilizes X-ray spectral analysis to investigate the properties of accreting black holes that extend over several orders of magnitude in accretion rate ($\dot{m}_{\rm E}$) and black hole mass ($M_{\rm BH}$), from the stellar mass black holes in X-ray binary systems (XRBs) to the supermassive black holes in active galactic nuclei (AGN). In general, high energy photons from accreting black holes are believed to originate from regions close to the event horizon where most of the gravitational potential energy is liberated, and emission in the X-ray band therefore provides the perfect probe for testing the geometry, behaviour and conditions present in the innermost regions of the accretion flow.

First, through a survey of X-ray emission in the nuclei of nearby galaxies I examine the usefulness of the X-ray to optical line ratios as identifiers of Compton-thick sources (i.e., sources with absorbing column densities greater than $\sim 10^{24}$atoms cm$^{-2}$) amongst low-luminosity AGN, and consider whether these ratios may have more practical use in distinguishing between AGN and non-AGN emission processes. Such linc ratios are commonly used to identify Compton thick sources amongst highly luminous AGN, but little is known about the extent to which this diagnostic applies to low-luminosity sources; the contribution to the optical line strengths from stellar processes in the latter may well be sufficient to affect the X-ray to optical line ratios, and therefore result in many sources being incorrectly classified as Compton thick.

Second, and more importantly, the main focus of this thesis is upon the variability of the Comptonised power-law X-ray spectral component, and more specifically an examination of how the photon index ($\Gamma$) varies with accretion rate. By cross-correlating the count rate from an XRB source with the fitted $\Gamma$ on time-scales as short as 16 ms I search for the presence of separate components which could be associated with Comptonisation of disc seed photons and Comptonisation of cyclo-synchrotron seed photons in the hot inner accretion flow.

I would like to express my sincere gratitude to my Ph.D. supervisor, Prof. Ian McHardy, for his infectious enthusiasm and limitless knowledge of all things in astrophysics, and for the many interesting chats over the years. I would also like to extend, thanks to the rest of the staff of the astronomy group (past and present), and especially Tom Maccarone and Phil Uttley, for their help and support.

My thanks also go to the Science and Technology Facilities Council (STFC) for financially supporting my research.

Finally, my deepest thanks go to my family, and this thesis would not have been written without their constant support and encouragement.

# Contents

# Figures

# Tables

# Chapter 1
# Introduction

*We are probably nearing the limit of all we can know about astronomy.*

Simon Newcomb (1888)

Although the first observational evidence for the existence of black holes in space was discovered only 50 years ago, much has been learnt in this time about the nature of these exotic objects and recognition of their importance in astrophysics has grown. Black holes, it seems, are the source of many contradictions. For instance, despite their gravitational attraction being so great that not even light can escape, it transpires that black holes are the most luminous objects in the Universe, and, additionally, are often responsible for driving the outflow of large quantities of matter at relativistic velocities! Perhaps most befittingly, their reputation as indiscriminate eating machines that consume all that venture too close is belied by their rôle in the formation and evolution of galaxies, in which they can justifiably be regarded as creators rather than destroyers. Despite our inability to directly image features such as an event horizon or an accretion disc, the extreme environments that exist in the vicinity of black holes provide a rich melting pot of interesting physics, and the lack of any direct imaging capability therefore seems only to increase the sense of mystery and further tantalise the curiosity.

In this chapter I present an overview of the various astronomical phenomena that are attributed to accretion onto black holes, and discuss how many of these can be unified under a model consisting of relatively few components. I briefly outline some of the different accretion mechanisms that may be responsible for the transport of matter towards the centre of a black hole, and discuss some of the physics that are required in order to understand them. In addition, I discuss the importance of the X-ray band to the study of black hole accretion, and describe the characteristics of a typical X-ray spectrum produced by this process. I also present an overview of the technical properties of the *Chandra X-ray Observatory* and *Rossi X-ray Timing*



C. Skipper, *Fast Spectral Variability in the X-ray Emission of Accreting Black Holes*, Springer Theses, DOI 10.1007/978-3-319-09587-5_1

*Explorer (RXTE)*, both of which have been used extensively throughout this thesis. Finally, I provide an outline of this thesis and discuss the main aims and objectives of this work.

## 1.1 Active Galactic Nuclei

### *1.1.1 Overview*

Active galactic nuclei (AGN) are compact regions at the centres of many galaxies from which the radiation output across the electromagnetic spectrum is much higher than can be explained by stellar processes alone. Galaxies that harbour AGN at their cores are often referred to as 'active galaxies', and are found at all redshifts and represent all morphological types.

The first sources to be recognised as being different from ordinary galactic nuclei were the Seyfert galaxies, first described by Carl Seyfert (Seyfert 1943). Seyfert recognised that many spiral galaxies exhibit broad Balmer emission lines in their optical spectra, and that the maximum width of these lines increases with the absolute magnitude of the nucleus.

Despite the identification of many more Seyfert galaxies over the following years, it wasn't until the discovery of large redshifts in the emission from quasi-stellar objects (quasars) in 1963 that astronomers first began to associate AGN activity with accretion onto a black hole. Quasars were originally identified from radio surveys in the late 1950s, and subsequently associated with sources appearing to be ordinary stars in optical images. However, their spectra revealed patterns of broad emission lines that could not initially be identified, and the proposition by Schmidt (1963) that these lines represented very highly redshifted Balmer series implied that the sources must be at very large distances and of unimaginable luminosity. At the time, the most important implication of this discovery concerned the big bang theory, as the lack of such highly luminous objects at small redshifts implied that the Universe had evolved over the last $10^9$ years so that the phenomena no longer existed. Today it is widely accepted that AGN must be powered by the release of gravitational potential energy as matter is accreted onto a supermassive black hole (SMBH; defined to be a black hole with mass greater than 100,000 $M_\odot$), and Seyfert galaxies and quasars simply represent examples of SMBH accretion at opposite ends of the luminosity scale.

In the early 1980s it was discovered that Seyfert galaxies are not the only class of low-luminosity AGN (LLAGN). Radio and optical surveys of the nuclei of bright galaxies (which, at the time, were believed to be ordinary galaxies) revealed that weak radio emission and optical emission lines were present in many of those studied (Heckman 1980). The spectra of these objects differed from those of Seyfert galaxies and quasars in that the emission lines appeared to originate from regions of much lower ionisation, and thereafter they were referred to as low-ionisation emission line regions (LINERs). In his original paper Heckman noted that LINERs appeared to occur more frequently in galaxies of early Hubble type, were often associated with compact nuclear radio sources, and exhibit emission lines that were of a similar width to those from narrow-line Seyferts, but lower in luminosity. Ho et al. (1997) later expanded upon this work with an extensive spectroscopic survey of 486 local galaxies, and found that LINERs make up the bulk of the AGN population in the local Universe, outnumbering Seyferts two to one. Ho also reported that the presence of LINERs strongly coincides with early-type hosts, while Seyfert nuclei are usually found (although not exclusively) in spirals.

The true nature of LINERs has often been the subject of debate. It has been suggested (e.g. Younes et al. 2011) that LINERs represent the AGN equivalent of hard-state X-ray binary systems, in which radiatively inefficient accretion flows (see below) are responsible for the X-ray emission. Alternatively, González-Martín et al. (2009) proposed that LINERs are heavily obscured AGN, with an inner absorber responsible for the X-ray absorption and an outer absorber responsible for the optical extinction. Other theories that have been suggested include post-starburst models (Taniguchi et al. 2000) and old stellar populations (Cid Fernandes et al. 2009), but the consensus of opinion has increasingly moved towards an AGN explanation for LINER emission.

### *1.1.2 The Unified Model*

Seyfert galaxies were originally classified into one of two types, depending upon whether their optical spectra contained both broad and narrow emission lines (type 1) or only narrow lines (type 2), and these were later extended to include a variety of sub-classes to reflect the relative strengths of the H $\alpha$ and H $\beta$ lines (Osterbrock 1977). In the 1980 s it was proposed by Antonucci and Miller (1985) (hereafter AM85) that the various classes of AGN can be unified within a single model in which the orientation of the system with respect to the observer is the main factor in producing the different observed properties (see Fig. 1.1). The gas producing the emission lines is ionised by the X-ray emission from the central engine, and the line widths are an indication of the Keplerian velocity of these gas clouds. It must be the case, therefore, that the broad line region (BLR) is considerably closer to the black hole than the narrow line region (NLR). The model of AM85 can only explain the data if we infer the existence of an absorber, which must take the form of a torus (although AM85 originally referred to this as an optically and geometrically thick disc), surrounding the central engine and obscuring the BLR from high-inclination (i.e. edge-on) observers.

The unified model of AM85 was inspired by their discovery of polarised broad emission lines in the optical spectrum of the Seyfert-2 galaxy NGC 1068, and similar discoveries have since been made in $\sim$50 % of Seyfert-2 galaxies for which we have suitable spectropolarimetric data. The presence of polarised broad lines implies that the type-2 source contains a hidden type-1 nucleus, the emission from which is

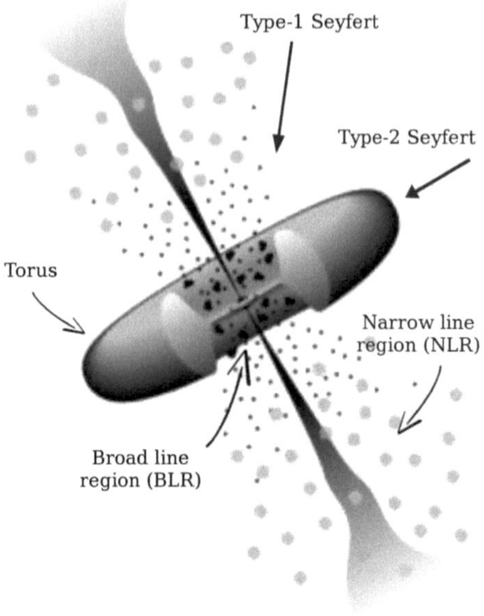


**Fig. 1.1** The AGN unification model, first proposed by Antonucci and Miller (1985). The BLR in the centre is obscured by the torus when the system is observed at a high inclination angle (i.e. edge-on), while the NLR further out is unaffected. This figure was modified from Urry and Padovani (1995), and is reproduced with permission of the University of Chicago Press

scattered into our line of sight by matter situated along the poles. In Chap. 2 I discuss the unified model in more detail, and look at some of the growing evidence of the model's inability to explain the X-ray spectra of a number of type-2 Seyferts.

## 1.2 X-ray Binaries

In 1962 the very first X-ray emission from outside our own solar system was detected by Geiger counters on board an Aerobee rocket (Giacconi et al. 1962). The source was eventually designated Scorpius X-1 and consists of a neutron star and a less massive stellar companion, V818 Scorpii, which orbits around it with a period of ~18.9 h. Since the early 60s many more examples of X-ray binaries (XRBs)—which consist of a compact object (black hole or neutron star) accreting matter from a secondary star—have been discovered, and the study of these has revealed a great deal of information about the physics of accretion. XRBs are divided into low-mass XRBs (LMXBs), where the donor star's mass is typically $<1\ M_\odot$, and high-mass XRBs (HMXBs), where the donor star's mass is $>1\ M_\odot$.

The transfer of matter from the donor star to the compact object can be either through the process of Roche-lobe overflow (which is always the case in LMXBs) or wind-fed accretion. In this thesis I refrain from discussing either of these processes in detail, and instead refer the reader to Paczyński (1971) and Mauche et al. (2008) for useful reviews. In this introduction I shall confine my discussion to consider only those processes that are responsible for accretion onto the compact object, and therefore the remainder of this chapter is relevant to both HMXBs and LMXBs, as well as AGN.

## 1.3 Mass Accretion onto Black Holes

### 1.3.1 Overview

Accretion, in astrophysics, is a term used to describe the steady flow of matter onto an object (which, throughout this thesis, is assumed to be a black hole). The large gravitational fields of compact objects are, on their own, insufficient to explain the high mass flow rates often detected from AGN and XRBs, and we therefore require accretion models which are able to explain how the accreting matter loses energy and angular momentum.

The total energy of a particle orbiting a black hole is given by the sum of the kinetic energy ($T$) and potential energy ($V$), such that

$$E = T + V = \frac{1}{2}mv^2 - \frac{GMm}{r}. \tag{1.1}$$

Differentiating the vector $\boldsymbol{r} = r\hat{\boldsymbol{i}}$ twice with respect to time (in a rotating coordinate system) gives

$$\dot{\boldsymbol{r}} = \dot{r}\hat{\boldsymbol{i}} + r\omega\hat{\boldsymbol{j}} \tag{1.2}$$

$$\ddot{\boldsymbol{r}} = (\ddot{r} - r\omega^2)\hat{\boldsymbol{i}} + \left(r\frac{d}{dt}(r^2\omega)\right)\hat{\boldsymbol{j}}, \tag{1.3}$$

where $\omega$ is the angular velocity.

The angular momentum is therefore

$$\boldsymbol{L} = m(\boldsymbol{r} \times \dot{\boldsymbol{r}}) = mr^2\omega\hat{\boldsymbol{k}}. \tag{1.4}$$

By relating Eq. 1.3 to the acceleration due to the gravitational force, $\ddot{\boldsymbol{r}} = -(GM/r^2)\hat{\boldsymbol{i}}$, and resolving in the $\hat{\boldsymbol{i}}$ and $\hat{\boldsymbol{j}}$ directions, we get

$$r\omega^2 - \ddot{r} = \frac{GM}{r^2}, \quad r\frac{d}{dt}(r^2\omega) = 0. \tag{1.5}$$

The second equation (the angular term) can be expressed as $r\dot{L}/m = 0$ (from Eq. 1.4) and confirms that angular momentum is conserved. Assuming circular motion (where $\ddot{r} = 0$ and $\omega = v/r$), the first equation simplifies to

$$v^2 = \frac{GM}{r}, \tag{1.6}$$

which can be substituted into the kinetic energy term of Eq. 1.1 to give

$$T = \frac{1}{2}\frac{GMm}{r}. \tag{1.7}$$

Differentiating $T$ and $V$ with respect to $r$ gives

$$\frac{dT}{dr} = -\frac{1}{2}\frac{GMm}{r^2}, \quad \frac{dV}{dr} = \frac{GMm}{r^2}, \tag{1.8}$$

or, in other words,

$$\frac{dT}{dr} = -\frac{1}{2}\frac{dV}{dr}. \tag{1.9}$$

This means that half of the potential energy liberated as accreting matter falls deeper into the potential well is converted into kinetic energy in order to maintain a circular orbit, and the other half must be lost either in the form of radiation or through some mechanical process (such as launching a jet).

The exact amount of gravitational potential energy that is liberated in the accretion process, and can therefore be radiated, depends upon the black hole spin, and is usually expressed in terms of the rest mass energy in the form $E_{\rm rad} = \eta mc^2$ (where $\eta$ is the efficiency). A simplistic approach would be to consider the innermost stable orbit of a Schwarzschild (non-rotating) black hole, given by $r_{\rm isco} = 6\,{\rm GM}/c^2$, at which a particle arriving from an infinite distance would have experienced a change in potential energy of $\Delta V = -mc^2/6$. Since half of this liberated energy would have been converted into kinetic energy then the maximum amount of energy that can be released as radiation is $mc^2/12$ (i.e. $\eta = 0.083$). This efficiency compares favourably with other forms of energy generation, such as nuclear fusion ($\eta \approx 0.003$), nuclear fission ($\eta \approx 0.001$) or burning coal ($\eta \approx 3 \times 10^{-11}$). For a Kerr (rotating) black hole the efficiency could possibly be as high as 0.4 (Shakura and Sunyaev 1973).

Since the energy radiated by the accretion process has to escape the system, and each photon carries a momentum $p_\gamma = E_\gamma/c$, then there must be an outward radiation pressure exerted upon the infalling matter. By summing over all the radiated photons, and then differentiating with respect to time, we get

$$\begin{aligned} \frac{d}{dt}\sum_\gamma p_\gamma &= \frac{1}{c}\frac{d}{dt}\sum_\gamma E_\gamma \\ \frac{dp}{dt} &= \frac{L_{\rm bol}}{c} \\ F_{\rm tot} &= \frac{L_{\rm bol}}{c} \end{aligned} \tag{1.10}$$

where $L_{\mathrm{bol}}$ is the bolometric (total) radiated luminosity. Assuming that the photons are radiated isotropically then the radiation pressure, $P_{\mathrm{rad}}$, at a distance $d$ becomes

$$P_{\mathrm{rad}} = \frac{L_{\mathrm{bol}}}{4\pi d^2 c}. \tag{1.11}$$

The radiation pressure will apply a force to all charged particles that is in proportion to the Thomson cross section for that particle, given by

$$\sigma_{\mathrm{T}} = \frac{8\pi}{3}\left(\frac{q^2}{4\pi\epsilon_0 m c^2}\right)^2. \tag{1.12}$$

Since $\sigma_{\mathrm{T}}$ is inversely proportional to $m^2$, and the mass of a proton is 1,836 times larger than that of an electron, then the Thomson cross section of an electron is 3.4 million times larger than that of a proton. If the infalling matter is comprised of hydrogen gas then the radiation pressure will therefore apply most strongly to the electrons, and the force on an atom will be

$$F_{\mathrm{rad}} = \frac{L_{\mathrm{bol}}\sigma_{\mathrm{T}}}{4\pi d^2 c}. \tag{1.13}$$

In contrast, the gravitational force on a hydrogen atom, $F_{\mathrm{G}} = GMm/d^2$, will apply most strongly to the proton. By equating $F_{\mathrm{rad}}$ and $F_{\mathrm{G}}$, and solving for $L_{\mathrm{bol}}$, we arrive at the Eddington limit (named after Arthur Eddington, who first proposed it)

$$L_{\mathrm{Edd}} = \frac{4\pi G M m_{\mathrm{p}} c}{\sigma_{\mathrm{T}}}, \tag{1.14}$$

and, in addition, the Eddington mass accretion rate

$$\dot{M}_{\mathrm{Edd}} = \frac{L_{\mathrm{Edd}}}{\eta c^2}. \tag{1.15}$$

The Eddington limit, although based upon a number of assumptions and simplifications (such as isotropic radiation and spherical accretion), provides a simple theoretical upper limit on the luminosity of an accretion-powered source. The ratio of the bolometric luminosity to the Eddington limit ($L_{\mathrm{bol}}/L_{\mathrm{Edd}}$, referred to as the Eddington ratio and usually denoted $\lambda$) is often used as a first order approximation to the accretion rate (which is itself denoted $\dot{m}_{\mathrm{E}}$ when expressed in Eddington units). A dimensionless form for the accretion rate is usually more useful than expressing this parameter using mass per unit time, as it enables a direct comparison between the accretion rates of multiple sources that may have very different black hole masses. However, it should be noted that the Eddington ratio is only an accurate proxy for $\dot{m}_{\mathrm{E}}$ at high accretion rates, where luminosity and accretion rate are related by $L_{\mathrm{bol}} \propto \dot{m}_{\mathrm{E}}$. When the source is accreting in a radiatively inefficient way then we instead expect $L_{\mathrm{bol}} \propto \dot{m}_{\mathrm{E}}^2$, and the Eddington ratio is no longer synonymous with the accretion rate.

### *1.3.2 Spherically Symmetric Accretion*

The simplest form of accretion is Bondi accretion (Bondi 1952), which describes the spherically symmetric accretion of a gas initially at rest at an infinite distance from the black hole. Bondi accretion was a theory based upon earlier works by Hoyle and Lyttleton (1939) and Bondi and Hoyle (1944), but simplified the problem through the assumption that the compact object was at rest with respect to the gas cloud. Under these conditions, the system will have no angular momentum and only the radial motion of the accreting gas needs to be considered. For spherically symmetric accretion, therefore, the infalling matter does not need to radiate its energy or lose angular momentum in order to accrete.

Bondi accretion was originally proposed to described the accretion of interstellar gas by stars and is not generally a realistic model for black hole accretion. The radiative efficiency of Bondi accretion is low and this theory is therefore not able to explain the large bolometric luminosities that are emitted by quasars, Seyferts, XRBs, etc., nor is it able to explain the large mass accretion rates that have been observed.

### *1.3.3 The Thin-Disc Accretion Model*

In order to explain the large luminosities that are often emitted by black hole accretion, we require an accretion mechanism that allows the gravitational potential energy to be radiated. In a seminal paper, Shakura and Sunyaev (1973) argued that matter flowing away from a star and onto its compact companion would have a large angular momentum, and would only be able to accrete if an effective mechanism existed for transporting this angular momentum outward. The authors proposed that the outward transfer of angular momentum leads to the formation of a disc (henceforth referred to as the accretion disc) around the black hole, and viscosity in this disc (defined to be $v = \alpha c_{\mathrm{s}} H$, where $\alpha$ is the viscosity parameter from Shakura and Sunyaev (1973), $c_{\mathrm{s}}$ is the sound speed and $H$ is the height of the disc) is responsible for the loss of part of the potential energy by radiation. The Shakura-Sunyaev disc is thin (with height $H$ much less than radius $R$), highly radiatively efficient and all motion in the disc is approximately Keplerian at all radii.

Material of mass $dM$ moving inwards through an annulus of the accretion disc of width $dr$ at radius $r$ will liberate potential energy

$$dV = \frac{GM}{r^2}\, dM\, dr. \tag{1.16}$$

From Eq. 1.9 we know that only half of this energy is radiated (with the remaining half converted into kinetic energy in order to maintain a circular orbit). The luminosity from this annulus is therefore

$$dL = \frac{1}{2}\frac{dV}{dt} = \frac{GM}{2r^2}\frac{dM}{dt}dr, \tag{1.17}$$

and this must be radiated from the top and bottom surfaces of the disc in accordance with the Stefan-Boltzmann law

$$dL = \epsilon\sigma T(r)^4 dA, \tag{1.18}$$

where $\epsilon$ is the emissivity, $\sigma$ is the Stefan-Boltzmann constant and $T$ is the radius-dependent temperature. The surface area of the annulus, $dA = 4\pi r dr$, includes an additional factor of 2 to represent the top and bottom surfaces. Rearranging gives

$$T(r)^4 = \frac{GM\dot{M}}{8\pi r^3 \epsilon\sigma}, \tag{1.19}$$

and expressing this equation in terms of the gravitational radius and Eddington accretion rate ($R_{\mathrm{G}} = GM/c^2$ and $\dot{M}_{\mathrm{Edd}} = L_{\mathrm{Edd}}/\eta c^2$ respectively, such that $r_{\mathrm{g}} = r/R_{\mathrm{G}}$ and $\dot{m}_{\mathrm{E}} = \dot{M}/\dot{M}_{\mathrm{Edd}}$) we obtain

$$T(r_{\mathrm{g}})^4 \propto \frac{1}{M}\frac{\dot{m}_E}{r_{\mathrm{g}}^3}. \tag{1.20}$$

The temperature therefore scales with $r_{\mathrm{g}}^{-3/4}$, and with $\dot{m}_{\mathrm{E}}^{1/4}$ and $M^{-1/4}$ at all radii. Consequently, for an equivalent accretion rate we would expect the accretion disc of a typical AGN with mass $10^7$ $\mathrm{M}_{\odot}$ to be ~30 times cooler than that of a typical XRB with mass 10 $\mathrm{M}_{\odot}$.

The luminosity of each small annuli is

$$dL = \frac{GM\dot{M}}{2r^2}dr, \tag{1.21}$$

and the total luminosity

$$L = \frac{GM\dot{M}}{2r_{\mathrm{in}}}. \tag{1.22}$$

The latter represents half of the potential energy liberated as material falls from an infinite distance to the inner edge of the disc (the truncation radius, $r_{\mathrm{in}}$). These equations show that most of the radiation is emitted from the innermost regions of the disc, and the truncation radius is equally as important as the accretion rate in determining the luminosity of the system.

### 1.3.4 Timescales in Accretion Disc Theory

There are several important time-scales which are relevant to accretion disc theory, and the simplest of these is the dynamical time-scale, $t_{\rm dyn}$, defined to be the Keplerian orbit time at radius $r$. From Eq. 1.6,

$$t_{\rm dyn} = 2\pi\sqrt{\frac{r^3}{GM}}. \tag{1.23}$$

The thermal timescale, $t_{\rm th}$, is the time required to establish thermal equilibrium in the disc, and is defined as the ratio of heat content in the disc to the dissipation rate. A simpler definition is in terms of the dynamical time-scale, such that

$$t_{\rm th} = \frac{t_{\rm dyn}}{\alpha}, \tag{1.24}$$

where $\alpha$ is the viscosity parameter.

Finally, I introduce the viscous time-scale, $t_{\rm visc}$, which is the time-scale on which matter diffuses through the disc, given by

$$t_{\rm visc} = \frac{t_{\rm th}}{(H/r)^2}, \tag{1.25}$$

where $H$ is the disc height. For typical accretion parameters we would find that

$$t_{\rm dyn} \ll t_{\rm th} \ll t_{\rm visc}. \tag{1.26}$$

### 1.3.5 Radiatively Inefficient Accretion Flows

In an attempt to explain the behaviour of many low accretion rate sources, a different set of accretion solutions were found (Shapiro et al. 1976; Ichimaru 1977) in which the accreting material is optically thin and radiatively inefficient. Accretion models that have these properties are known as radiatively inefficient accretion flows (RIAFs) and consist of a two-temperature plasma in which the ions and electrons are no longer in thermal equilibrium (e.g. Narayan 1996). The ions, being poor radiators, become significantly hotter than the electrons ($T_{\rm i}{\sim}10^{11}$ K, $T_{\rm e}{\sim}10^8$–$10^9$ K; Narayan et al. 1998) and the accretion flow becomes geometrically thick (puffed up). The best known example of a RIAF model is the advection dominated accretion flow (ADAF;

e.g. Narayan and Yi 1994) in which the accretion energy is advected through the event horizon.

In ADAF theory there exists a critical accretion rate ($\dot{m}_{\rm crit}$) above which the existence of an ADAF is not possible. This value, measured in Eddington units, is a function of radius such that (from Narayan et al. 1998)

$$\dot{m}_{\rm crit} \propto \frac{\alpha^2}{\sqrt{r}}, \tag{1.27}$$

where $\alpha$ is the viscosity parameter. The implication of this relation is that, at a certain range of accretion rates, an ADAF can exist only at low values of $r$. This scenario, in which an inner RIAF gives way to an outer thin disc (e.g. Esin et al. 1997; Done and Gierliński 2006; Qiao and Liu 2010), is referred to as the two-zone accretion or truncated disc model and largely forms the basis of all discussion and interpretation within this thesis. The radius at which the outer disc evaporates into an inner hot, thick flow is called the truncation radius and is denoted $r_{\rm in}$.

Narayan et al. (1998) expressed the energy conservation equation of black hole accretion in the form $q^{\rm adv} = q^+ - q^-$, where $q^{\rm adv}$ is the advective transport of energy, $q^+$ is the energy generated by viscosity and $q^-$ is the radiative cooling. The three forms of accretion mechanism discussed in this section can therefore be summarised by the conditions $q^+ \approx q^- \gg q^{\rm adv}$ (thin disc accretion, where potential energy is liberated by friction in the disc and efficiently radiated), $q^{\rm adv} \approx q^+ \gg q^-$ (RIAFs, where viscous energy is carried into the black hole by the accretion flow) and $-q^{\rm adv} \approx q^- \gg q^+$ (Bondi accretion, where some of the thermal energy in the accretion flow is radiated and almost no energy is generated by viscosity).

## 1.4 The X-ray Spectra of Accreting Black Holes

### *1.4.1 Overview*

The energy radiated by the accretion process is distributed right across the electromagnetic spectrum, but in this thesis I shall focus predominantly on the X-ray emission. High energy photons contribute greatly to the overall bolometric luminosity of an accreting black hole, and the origin of this emission must be the innermost region of the accretion flow where temperatures are high enough and the gravitational and magnetic forces strong enough to accelerate particles to relativistic velocities. Although X-ray observatories have insufficient angular resolution to resolve features on these very small scales, spectral and timing analysis can reveal a great deal about the physical processes taking place in these extreme environments.

The X-ray emission from accreting black holes usually consists of just a few common components, all of which are though to originate in two regions of the accretion flow. The first is the accretion disc, which contributes black body emission to the

X-ray band in XRBs but not AGN (AGN accretion discs are much cooler and typically radiate in the UV—see Eq. 1.20), along with reprocessed emission in the form of reflection and an iron K $\alpha$ fluorescence line at $\sim$6.4 keV. In addition to the accretion disc, we also find X-ray emission due to inverse-Compton scattering of seed photons by hot electrons in the vicinity of the black hole. However, the exact nature of the Comptonising electrons is not always clear. For instance, at low accretion rates when the disc is truncated (referred to as the hard state) then the hot, geometrically thick inner accretion flow can provide the relativistic electrons required for Comptonisation. Alternatively, the hard state in XRBs has been associated with strong radio emission, which indicates the presence of powerful jets (relativistic outflows of charged particles along the system's angular momentum vector), and such jets could perhaps be the source of Comptonisation. However, at higher accretion rates a thin disc often extends all the way down to the innermost stable orbit (referred to as the soft state) and strong jets and RIAFs are not present. Instead, the Comptonising electrons in the soft state are thought to take the form of either a corona or a disc wind, and are likely to be found above the innermost regions of the accretion disc. For the purposes of this thesis, and by convention, I shall often refer to the relativistic electrons (including those in a geometrically thick flow or in the jets) simply as "the corona" without implying or favouring any particular geometry over another.

## *1.4.2 Inverse-Compton Scattering*

Inverse-Compton scattering (or simply Comptonisation) is the process whereby a photon scatters off a high energy electron such that the photon will gain energy from the collision and the electron will lose energy. Figure 1.2 represents an interaction between a photon and an electron with initial momentum $p^{\mu}_{\gamma \mathrm{i}}$ and $p^{\mu}_{\mathrm{ei}}$ respectively, and final momentum $p^{\mu}_{\gamma \mathrm{f}}$ and $p^{\mu}_{\mathrm{ef}}$ respectively, such that

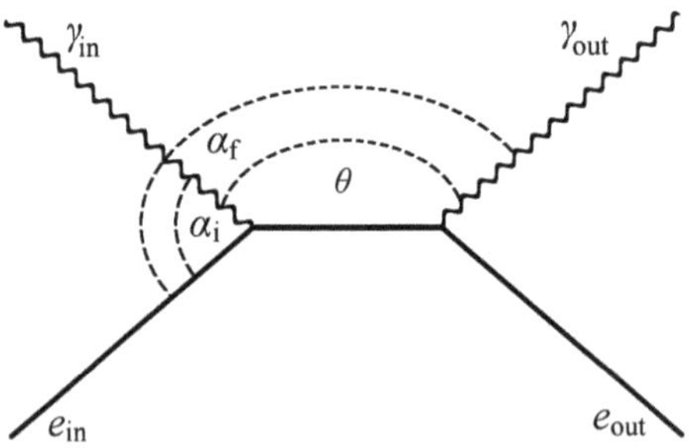


**Fig. 1.2** An interaction between a photon with initial energy $h\nu_{\mathrm{i}}$ and an electron with initial energy $\gamma_{\mathrm{i}} m_{\mathrm{e}} c^2$. The photon scatters through an angle $\theta$

$$p_{\gamma\mathrm{i}}^{\mu} = \frac{h\nu_{\mathrm{i}}}{c}\begin{pmatrix}1\\ \hat{\boldsymbol{n}}_{\gamma\mathbf{i}}\end{pmatrix} \qquad p_{\gamma\mathrm{f}}^{\mu} = \frac{h\nu_{\mathrm{f}}}{c}\begin{pmatrix}1\\ \hat{\boldsymbol{n}}_{\gamma\mathbf{f}}\end{pmatrix}$$
$$p_{\mathrm{ei}}^{\mu} = \gamma_{\mathrm{i}} m_{\mathrm{e}}\begin{pmatrix}c\\ v_{\mathrm{i}}\hat{\boldsymbol{n}}_{\mathbf{ei}}\end{pmatrix} \qquad p_{\mathrm{ef}}^{\mu} = \gamma_{\mathrm{f}} m_{\mathrm{e}}\begin{pmatrix}c\\ v_{\mathrm{f}}\hat{\boldsymbol{n}}_{\mathbf{ef}}\end{pmatrix} \tag{1.28}$$

where $\hat{\boldsymbol{n}}_{\gamma\mathbf{i}}$, $\hat{\boldsymbol{n}}_{\mathbf{ei}}$, $\hat{\boldsymbol{n}}_{\gamma\mathbf{f}}$ and $\hat{\boldsymbol{n}}_{\mathbf{ef}}$ represent the initial and final unit directional vectors of the photon and electron, and $v_{\mathrm{i}}$ and $v_{\mathrm{f}}$ represent the initial and final velocities of the electron.

Applying conservation of energy and momentum gives

$$p_{\gamma\mathrm{i}}^{\mu} + p_{\mathrm{ei}}^{\mu} = p_{\gamma\mathrm{f}}^{\mu} + p_{\mathrm{ef}}^{\mu}, \tag{1.29}$$

which I choose to express as

$$p_{\mathrm{ef}}^{\mu} = p_{\gamma\mathrm{i}}^{\mu} + p_{\mathrm{ei}}^{\mu} - p_{\gamma\mathrm{f}}^{\mu}. \tag{1.30}$$

The Lorentz invariant product is

$$\begin{gathered} p_{\mathrm{ef}}^{\mu}\, p_{\mathrm{ef}\,\mu} = m_{\mathrm{e}}c^2 \\ \left(p_{\gamma\mathrm{i}}^{\mu} + p_{\mathrm{ei}}^{\mu} - p_{\gamma\mathrm{f}}^{\mu}\right)\left(p_{\gamma\mathrm{i}\,\mu} + p_{\mathrm{ei}\,\mu} - p_{\gamma\mathrm{f}\,\mu}\right) = m_{\mathrm{e}}c^2, \end{gathered} \tag{1.31}$$

which can be simplified using the Lorentz invariant products $p_{\mathrm{ei}}^{\mu}\, p_{\mathrm{ei}\,\mu} = m_{\mathrm{e}}c^2$, $p_{\gamma\mathrm{i}}^{\mu}\, p_{\gamma\mathrm{i}\,\mu} = 0$ and $p_{\gamma\mathrm{f}}^{\mu}\, p_{\gamma\mathrm{f}\,\mu} = 0$ to give

$$\nu_{\mathrm{i}}\left(c - v_{\mathrm{i}}\hat{\boldsymbol{n}}_{\mathbf{ei}}\cdot\hat{\boldsymbol{n}}_{\gamma\mathbf{i}}\right) - \nu_{\mathrm{f}}\left(c - v_{\mathrm{i}}\hat{\boldsymbol{n}}_{\mathbf{ei}}\cdot\hat{\boldsymbol{n}}_{\gamma\mathbf{f}}\right) - \frac{h\nu_{\mathrm{i}}\nu_{\mathrm{f}}}{\gamma_{\mathrm{i}} m_{\mathrm{e}}c^2}\left(1 - \hat{\boldsymbol{n}}_{\gamma\mathbf{i}}\cdot\hat{\boldsymbol{n}}_{\gamma\mathbf{f}}\right) = 0, \tag{1.32}$$

or

$$\lambda_{\mathrm{f}}\left(1 - \beta_{\mathrm{i}}\cos\alpha_{\mathrm{i}}\right) - \lambda_{\mathrm{i}}\left(1 - \beta_{\mathrm{i}}\cos\alpha_{\mathrm{f}}\right) = \frac{h}{\gamma_{\mathrm{i}} m_{\mathrm{e}}c^2}\left(1 - \cos\theta\right), \tag{1.33}$$

where $\beta_{\mathrm{i}} = v_{\mathrm{i}}/c$.

For normal Compton scattering we would regard the electron as being initially at rest (so $\beta_{\mathrm{i}} = 0$ and $\gamma_{\mathrm{i}} = 1$), and would recover the familiar equation

$$\lambda_{\mathrm{f}} - \lambda_{\mathrm{i}} = \frac{h}{m_{\mathrm{e}}c^2}\left(1 - \cos\theta\right), \tag{1.34}$$

but for inverse-Compton scattering the electron is relativistic and Eq. 1.33 cannot be simplified much further.

Instead, the problem can be tackled slightly differently by making some additional assumptions. Firstly, the scattering can be assumed to occur in one dimension so that the various angles need not be considered, and, secondly, the electron momentum can be assumed to be much greater than the seed photon momentum ($\gamma m_{\mathrm{e}} v \gg h\nu_{\mathrm{i}}/c$)

so that the photon scatters through 180° and the electron energy and momentum is considered to be unaffected (which is, essentially, simplifying the problem by conserving energy but not momentum; in addition, the subscripts $i$ and $f$ have now been dropped from $\gamma$ and $v$ since $\gamma_{\rm i} \approx \gamma_{\rm f}$ and $v_{\rm i} \approx v_{\rm f}$). If the photon is initially travelling in the $-x$ direction with energy $h\nu_{\rm i}$ and the electron is travelling in the $+x$ direction with speed $v$ then we can boost the photon into the rest frame of the electron

$$\begin{aligned} p^{\mu'}_{\gamma \rm i} &= \begin{pmatrix} \gamma & -v/c \\ -v/c & \gamma \end{pmatrix} \begin{pmatrix} h\nu_{\rm i}/c \\ -h\nu_{\rm i}/c \end{pmatrix} \\ p^{\mu'}_{\gamma \rm i} &= \frac{h\nu_{\rm i}}{c} \begin{pmatrix} \gamma + v/c \\ -\gamma - v/c \end{pmatrix}. \end{aligned} \tag{1.35}$$

The electron is scattered through 180° so that the sign of its momentum changes, and we can then boost back into the original frame

$$\begin{aligned} p^{\mu}_{\gamma \rm f} &= \begin{pmatrix} \gamma & v/c \\ v/c & \gamma \end{pmatrix} \frac{h\nu_{\rm i}}{c} \begin{pmatrix} \gamma + v/c \\ \gamma + v/c \end{pmatrix} \\ p^{\mu}_{\gamma \rm f} = \begin{pmatrix} h\nu_{\rm f}/c \\ h\nu_{\rm f}/c \end{pmatrix} &= \frac{h\nu_{\rm i}}{c} \begin{pmatrix} \gamma^2 + 2\gamma v/c + v^2/c^2 \\ \gamma^2 + 2\gamma v/c + v^2/c^2 \end{pmatrix}. \end{aligned} \tag{1.36}$$

The energy of the photon is therefore boosted in the scattering by a factor of $\gamma^2 + 2\gamma v/c + v^2/c^2$ or $(\gamma + v/c)^2$. Since the electron is relativistic, and $v < c$, then $\gamma \gg v/c$ and we can say that

$$h\nu_{\rm f} \approx \gamma^2 h\nu_{\rm i}. \tag{1.37}$$

A black body disc spectrum that is Compton up-scattered by electrons of energy $\gamma m_{\rm e} c^2$ will have a Comptonised spectrum that is of the same shape, but shifted by a factor of $\gamma^2$ to higher energies. However, the electrons will not all be of the same energy and will instead have their own energy spectrum so that the Comptonised spectrum will be the convolution of the electron and seed photon energy spectra. In addition, photons can be Compton scattered multiple times before escaping the electron cloud so the final Comptonised spectrum can be approximated as the sum of many Compton up-scattered black body spectra (see Fig. 1.3).

Inverse-Compton scattered spectra typically take the form of a power-law (where $F_\nu \propto E^{-\Gamma}$ and $\Gamma$ is called the photon index), but this power-law often breaks (turns down) at low and/or high energies. The lower energy break will occur at energies similar to the typical seed photon energy, and can therefore be used to distinguish between disc black body emission and cyclo-synchrotron emission as the main source of seed photon supply. If the Comptonising electrons are thermally distributed (thermal Comptonisation) then we would also expect to find a break in the spectrum at high energies as a seed photon cannot be up-scattered to an energy higher than that available to the scattering electrons. Thermal Comptonisation is usually successful in describing the spectra of hard-state XRBs, although the source of seed photons can vary depending upon the accretion rate. The alternative to thermal

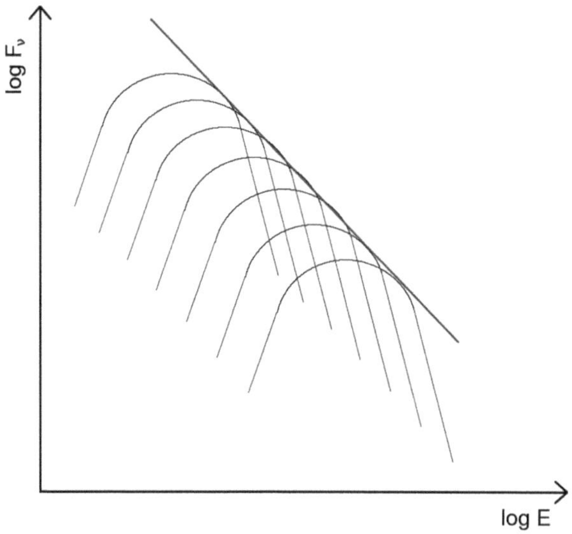


**Fig. 1.3** Inverse-Compton scattering of a black body seed spectrum. The sum of many scatterings takes the form of a power law, $F_\nu \propto E^{-\Gamma}$

Comptonisation is non-thermal Comptonisation, in which the scattering electrons have a power-law energy distribution ($n(\gamma) \propto \gamma^{-p}$) and the power-law Comptonised spectrum can extend out to very large energies (>800 keV for Cygnus X-1, see Gierliński et al. 1999). Non-thermal Comptonisation is usually applied to the soft state of XRBs, suggesting a switch in the electron energy distribution during a state transition.

### 1.4.3 Spectral States in X-ray Binaries

The broadband spectra of black hole XRBs can usually be divided into characteristic states, defined by their spectral and timing properties. The exact number of states can be fairly arbitrary—and their defining properties have often been refined over the years—but can roughly be described as the quiescent state, the hard (or low) state (HS), the intermediate state (IS, usually divided into hard- and soft-intermediate state), the soft (or high) state (SS) and the very high state (VHS). Figure 1.4 shows a simple diagram of the soft and hard states, with the former represented by a thin accretion disc at all radii beyond the innermost stable orbit and a compact corona above the inner region of the disc. The exact geometry of the corona is not known, and may instead take the form of a disc wind; alternatively, Churazov et al. (2001) proposed that the thin disc in both the soft and hard states is sandwiched by the corona out to a large distance from the black hole. The truncated thin disc in the hard state is replaced at smaller radii by a geometrically thick, optically thin and

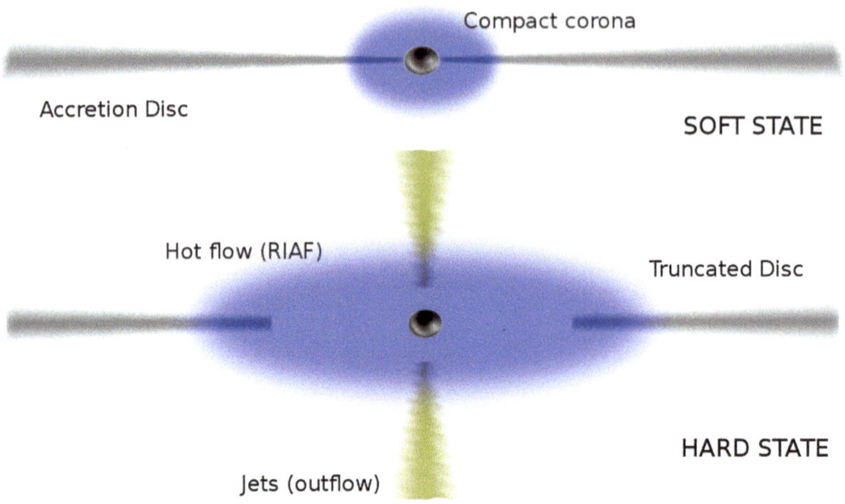


**Fig. 1.4** Diagram of the accretion flow in the soft and hard states. The exact geometry of the soft-state corona is unclear, and it may instead take the form of a disc wind. The truncated thin disc in the hard state is replaced at smaller radii by a geometrically thick, optically thin and radiatively inefficient accretion flow (RIAF)

radiatively inefficient accretion flow (RIAF) which launches powerful relativistic jets along the polar axes. Whilst in the hard state, XRBs are usually found to have relatively stronger radio emission than in the soft state, and greater fractional RMS variability.

One of the best ways of distinguishing these states is by plotting the hardness ratio (the ratio of count rate or flux in two separate X-ray bands) against the source count rate, which is usually referred to as the hardness-intensity diagram. An example of such a plot is shown in Fig. 1.5, which shows the evolution of the low-mass XRB GX 339-4 during its 2002/2003 outburst.

When plotted on the hardness-intensity diagram the data from GX 339-4 take the form of a q-shaped hysteresis curve (called the turtle-head diagram), which is always traversed in an anti-clockwise direction. The arrows on the plot (starting from the upwards arrow on the right-hand side) represent the hard state, the intermediate state, the soft state and the intermediate state respectively. GX 339-4 is amongst a number of sources that enter and leave the soft state at different accretion rates (e.g. XTE J1650-500, XTE J1859+226; see Belloni 2010), but this behaviour is not common to all black hole XRBs (Cygnus X-1 being one notable exception). There will be further discussion of the hardness-intensity diagram in Chap. 4.

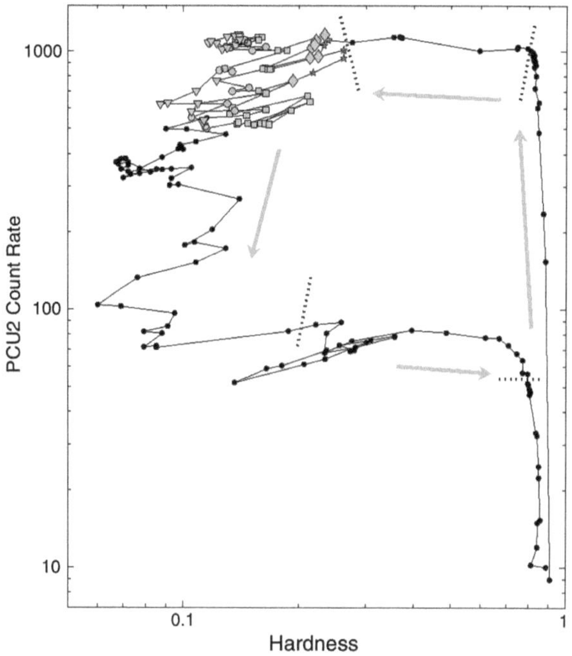


**Fig. 1.5** The hardness-intensity diagram of GX 339-4 during its 2002/2003 outburst (Belloni et al. 2005). The data form a characteristic *q-shaped* hysteresis curve (called the turtle-head diagram) which is always traversed in an anti-clockwise direction. Reproduced with permission © ESO

## *1.4.4 The X-ray Spectral Components of XRBs and AGN*

In this section I have already discussed the Comptonised power-law and disc black body emission often found in the spectra of accreting black holes. In addition to these components it is often necessary to model photoelectric absorption and reprocessed emission such as reflection and an iron K $\alpha$ fluorescence line (see Fig. 1.6). Photoelectric absorption is the process by which an incoming photon transfers all its energy to an atomic electron (usually a K-shell electron) resulting in the electron's ejection from the atom. For an observer viewing an X-ray source through an absorbing cloud of gas, the loss of flux due to photoelectric absorption will be energy dependent, and takes the form

$$n(E) = n_0(E)\,\mathrm{e}^{-N_{\mathrm{H}}\sigma(E)}, \tag{1.38}$$

where $n(E)$ is the observed photon number density, $n_0(E)$ is the intrinsic (unabsorbed) photon number density, $N_{\mathrm{H}}$ is the column density of hydrogen gas (atoms per cm$^2$) and $\sigma(E)$ is the photo-electric cross section (not including Thomson scattering). The photo-electric cross section is greater at soft X-ray energies, and absorp-

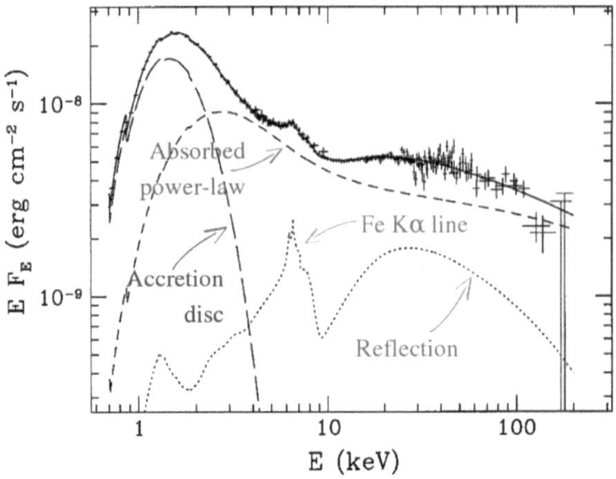


**Fig. 1.6** X-ray spectra of the HMXB Cygnus X-1, showing an accretion disc, Comptonised power-law (mildly absorbed at low energies) and reflection and iron line components (Gierliński et al. 1999). Reproduced with permission of Oxford University Press on behalf of the Royal Astronomical Society

tion is therefore experienced more strongly at low energies than at high energies (as demonstrated by Fig. 1.6).

The reflection component represents reprocessed emission from the Comptonised power-law and is characterised by a prominent hump at ~30 keV and a strong iron $K\alpha$ fluorescence line at ~6.4 keV. The reprocessor is usually believed to be either the accretion disc or, for AGN, the molecular torus that surrounds the system, and the reflection is most strongly observed above ~10 keV where it causes a typical hardening of the spectra. Reflection can be modelled in a variety of ways but, in this thesis, I choose to use the models of Magdziarz and Zdziarski (1995) which apply to both ionised (PEXRIV) and non-ionised (PEXRAV) material. The iron line is the result of excitation of atoms in the accretion disc (or, for AGN, narrow iron lines may also originate in the molecular torus) by Comptonised photons from the corona, and often displays extreme relativistic effects. Broadened iron lines are generally indicative of reprocessing from the innermost regions of the accretion disc, but gravitational redshift and Compton scattering reduce the symmetrical appearance of the line and further complicate the analysis.

The soft-state 0.3–20 keV X-ray spectra of XRBs typically show a strong thermal component from the accretion disc which can dominate energies up to ~10 keV, a non-thermal Comptonised power-law tail (with photon index, $\Gamma$, of between 2 and 3) and strong iron line and reflection components. The hard-state spectra in this energy range are usually much simpler, and are dominated by a strong thermal Comptonised power-law (with $\Gamma$ roughly in the range 1.4–1.8) complemented by weak reflection and iron line components. As this thesis does not examine the spectra at energies higher than 20 keV, I need not be too concerned with the high energy break in the power-law.

The X-ray spectra of AGN differ from those of XRBs in a number of ways. Firstly, the disc black body component is not present at X-ray energies in AGN, but is instead replaced in some sources by a much more mysterious "soft excess". The soft excess has generally been found to be a poor fit to disc black body models, and attempts to do so find the temperature to be constant across many sources and accretion rates (e.g. Gierliński and Done 2004). Secondly, although AGN show reflection from the accretion disc there may also be a significant reprocessed component originating from the inner edge of the torus. The unified model asserts that in many type-2 AGN the orientation of the torus is such that the emission from the innermost regions (i.e. the intrinsic black body and power-law emission from the disc and corona respectively) are completely suppressed, and instead we see a strong reflection spectrum, presumably from the inner edge of the far side of the torus. Studies of classical type-2 AGN, such as NGC 1068, generally support this theory (e.g. Iwasawa et al. 1997; Matt et al. 2001) and find that most of the iron line flux is from a narrow component emitted from cold, neutral matter situated at a larger distance from the black hole than can be attributed to the accretion disc (i.e. the torus, perhaps); the broad component, emitted from warm, ionised matter close to the black hole is, by comparison, relatively weak.

### *1.4.5 Variability in Black Hole Accretion*

The X-ray emission from XRBs and AGN is known to vary on all time-scales, and, in the soft state of XRBs, most of this variability is associated with the power-law component (Churazov et al. 2001). Uttley and McHardy (2001) found that the rms variability is linearly related to the average flux (the rms-flux relation), which implies that the source of the variability is in the accretion flow. When combined with the presence of hard lags in the data (whereby the hard X-rays are found to slightly lag the soft X-rays—e.g. see Kotov et al. 2001) the rms-flux relation favours the model of Lyubarskii (1997) in which fluctuations in the accretion rate propagate inwards through the accretion flow and, as they do so, are continuously modulated by more rapid local accretion rate fluctuations. To produce a hard lag requires that the emission region is both extended and radially stratified, such that the more energetic (hotter) Comptonising electrons are, on average, found closer to the black hole than the cooler electrons. Under this scenario, both the hard energy photons and the most rapid variability are produced in the innermost regions of the accretion flow.

In this thesis I focus mainly upon the variability of the power-law spectral component, which is ubiquitous in the X-ray spectra of both XRBs and AGN and is known to vary on a wide range of time-scales. Previous studies of black hole X-ray spectra have established a correlation, within samples of AGN, between the power-law photon index ($\Gamma$) and the source accretion rate ($\dot{m}_{\rm E}$) (e.g. Constantin et al. 2009; Gu and Cao 2009; Shemmer et al. 2006), and a similar relationship has also been found within individual XRBs (e.g. Wu and Gu 2008; Sobolewska et al. 2011; Zdziarski et al. 2002). Above a critical accretion rate ($\dot{m}_{\rm crit}$), typically 0.5–1 % of the Eddington limit, $\dot{m}_{\rm edd}$, $\Gamma$ is found to be positively correlated with the accretion rate (the source is softer when brighter) while below $\dot{m}_{\rm crit}$ these two properties are anti-correlated (the source is harder when brighter). This switch in the relationship between $\dot{m}_{\rm E}$ and $\Gamma$ is not the same as the state transition from the hard to the soft state, which usually occurs at a higher accretion rate of around a few percent of the Eddington limit or greater (e.g. Maccarone 2003; Gierliński and Newton 2006).

Within individual AGN the positive $\Gamma$-$\dot{m}_{\rm E}$ correlation is well established on time-scales of months to years (e.g. Lamer et al. 2003; Sobolewska and Papadakis 2009) as well as on very short time-scales of hours (Ponti et al. 2006). The negative correlation has also been found in one AGN on long time-scales (Emmanoulopoulos et al. 2012). Within individual XRBs both the positive and negative correlations have been observed on long time-scales of days to years (Wu and Gu 2008; Sobolewska et al. 2011; Zdziarski et al. 2002), and, although little is known about the relationship between $\Gamma$ and $\dot{m}_{\rm E}$ on very short time-scales, Wu et al. (2010) showed that the positive correlation exists on time-scales of 62 ms in a single observation of Cygnus X-1.

## 1.5 The Chandra X-ray Observatory

The *Chandra X-ray Observatory* is a NASA mission that was launched on 1999 July 23 into an elliptical high-earth orbit with a period of ~64 h. The telescope consists of four pairs of highly-polished mirrors and four scientific instruments: the High Resolution Camera (HRC), the Advanced CCD Imaging Spectrometer (ACIS), the High-Energy Transmission Grating Spectrometer (HETGS) and Low-Energy Transmission Grating Spectrometer (LETGS); see Fig. 1.7 for a diagram of the telescope layout.

The HRC is an imaging instrument (consisting of two micro-channel plates) sensitive to the 0.1–10 keV energy range, and has better than $0.5''$ angular resolution. The ACIS consists of ten Charge-Coupled Device (CCD) chips split into four ACIS-I chips (arranged in a $2 \times 2$ grid, numbered I0–I3, and used specifically for imaging) and six ACIS-S chips (arranged in a $1 \times 6$ array, numbered S0–S5, and used either for imaging or as a grating readout). Each CCD offers $1024 \times 1024$ image resolution and targets are usually aimed onto either the I3 or S3 CCDs. When the ACIS is used in combination with either the LETGS or HETGS then *Chandra* is able to provide excellent energy resolution of up to $E/dE \approx$ 1,000. In addition, ACIS provides an angular resolution that is better than $0.5''$. In this thesis I have used the ACIS instrument for studying AGN and, where required, I have also used the ACIS in combination with either the LETGS or HETGS. Aside from the improved spectral resolution that is obtained when using one of the transmission gratings, the sensitivity is also somewhat reduced which can help to reduce the effect of pileup for the brighter sources.

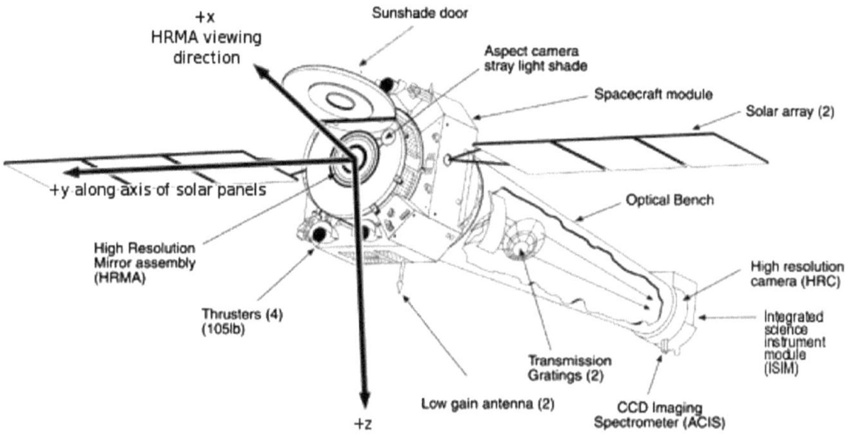


**Fig. 1.7** Diagram of the *Chandra* X-ray observatory showing the location of the CCD Imaging Spectrometer (ACIS). This figure is from the *Chandra* website (Credit: NASA/CXC/NGST)

## 1.6 The Rossi X-ray Timing Explorer

The Rossi X-ray Timing Explorer (*RXTE*) was a NASA mission that was launched on 1995 December 30 into a circular low-earth orbit with a period of 90 min. Despite an initial goal of a 5 year mission time, the spacecraft continued to operate successfully for 16 years. *RXTE* is unusual amongst X-ray observatories of the time in that it has no imaging capability and is instead designed to study X-ray variability on time-scales as short as 1 $\mu$s.

*RXTE* carries three instruments: the Proportional Counter Array (PCA), the High Energy X-ray Timing Experiment (HEXTE) and the All-Sky Monitor (ASM); see Fig. 1.8 for a diagram of the observatory layout. The PCA and HEXTE are pointing instruments, and are collimated to provide a field of view of 1°, whereas the ASM is a wide-field instrument and scans 80 % of the sky (in the 2–10 keV range) with every orbit. The PCA consists of an array of five proportional counters (numbered 0–4) and covers the energy range 2–60 keV. The PCA's timing resolution of 1 $\mu$s is unprecedented, although the spectral resolution of $dE/E \approx 0.18$ at 6 keV is only moderate. The PCA data are provided in either event mode or binned mode, with the former recording each individual photon event and the latter being binned into a reduced number of energy and time bins. The simplest and most commonly used data mode is the Standard-2 binned format, which provides maximum spectral resolution and 16 s time-resolution; the Standard-2 data format is normally the mode of choice for observers unless better time resolution is required. The HEXTE instrument consists of two clusters of scintillation counters covering the energy range 15–250 keV. Although the HEXTE time resolution of 8 $\mu$s is a little lower than that of the PCA, the spectral resolution of $dE/E \approx 0.15$ at 60 keV is slightly better.

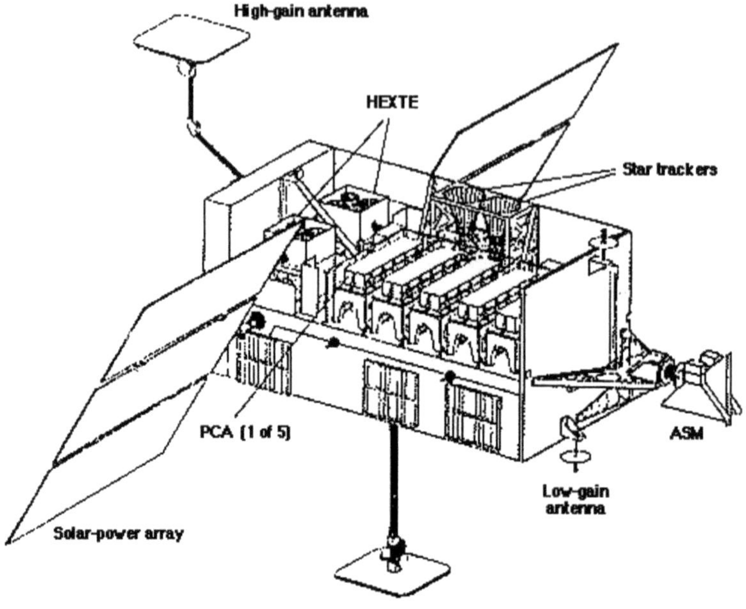


**Fig. 1.8** Diagram of the *Rossi X-ray Timing Explorer*, showing the location of the main instruments. This figure is from the HEASARC website (Credit: NASA)

With no imaging capability to allow background regions to be selected, the background subtraction process in RXTE data reduction instead relies upon a model designed to estimate not only the X-ray background but also the effect of cosmic ray bombardment and other particle contamination. At high energies, where the X-ray background begins to dominate over the source, the reliability of the spectra from many sources can become reduced. In this thesis I therefore use only the PCA instrument to examine energies up to 20 keV.

## 1.7 Summary of This Thesis

The main focus of this thesis is upon two separate, but loosely related, aspects of black hole research through the X-ray waveband that ultimately combine to allow a comparison to be drawn between the behaviour of XRBs and AGN. Firstly, I examine the properties of nearby galactic nuclei through an unbiased *Chandra* snapshot survey which I complement with data from other wavelengths that have been harvested from the literature. This work was originally undertaken as part of a project (the LeMMINGS survey) expected to provide e-MERLIN radio data on a sample of nearby galaxies, but unfortunately the delivery of these data was significantly delayed and the focus of my research therefore had to change. Nevertheless, in Chap. 2 I present the results from my X-ray survey and discuss the effect of absorption on the data. In particular, I examine whether the often-used relationship between the X-ray luminosity and the strength of various optical emission lines (such as [O III], H $\alpha$, etc.) is a suitable diagnostic of Compton thickness, and discuss what these results could mean to our understanding of the emission processes in LLAGN. Although this chapter may seem a little disconnected from the variability work that follows, it nevertheless provides some relevant information on the same sample of LLAGN and the extent to which absorption may be affecting their X-ray observations

The second, and most important, focus of this thesis is upon the variability of the X-ray power-law component in both XRBs and AGN, on time-scales that allow the behaviour of these two classes of object to be compared. The X-ray photon index, $\Gamma$, is known to be correlated with the source flux in a variety of different ways that depend upon the spectral state and accretion rate, and in Chap. 3 I examine the relationship between the source count rate and $\Gamma$ for the HMXB Cygnus X-1 by extracting lightcurves of count rate and $\Gamma$ from *RXTE* data on millisecond time-scales and cross-correlating these properties.

In Chap. 4 I introduce a new technique for examining the $\dot{m}_{\rm E}$–$\Gamma$ relationship in XRBs that involves many spectra with similar count rates (extracted from *RXTE* data with a 100 ms timing resolution) being 'summed' to produce a single spectrum with improved signal-to-noise ratio. This technique is also applied to AGN, for which the black hole mass/variability time-scale scaling relationship requires the spectral timing resolution to be of the order of a few days. I thereby examine the $\dot{m}_{\rm E}$–$\Gamma$ distribution plots of XRBs and AGN on comparable (mass-scaled) time-scales.

## References

Antonucci, R. R. J., & Miller, J. S. (1985). Spectropolarimetry and the nature of NGC 1068. *The Astrophysical Journal*, *297*, 621–632.

Belloni, T., Homan, J., Casella, P., van der Klis, M., Nespoli, E., Lewin, W. H. G., et al. (2005). The evolution of the timing properties of the black-hole transient GX 339–4 during its 2002/2003 outburst. *Astronomy and Astrophysics*, *440*, 207–222.

Belloni, T. M. (2010). States and transitions in black hole binaries. In T. Belloni (Ed.), Lecture Notes in Physics, vol. 794 (p. 53). Berlin: Springer

Bondi, H. (1952). On spherically symmetrical accretion. *Monthly Notices of the Royal Astronomical Society*, *112*, 195.

Bondi, H., & Hoyle, F. (1944). On the mechanism of accretion by stars. *Monthly Notices of the Royal Astronomical Society*, *104*, 273.

Churazov, E., Gilfanov, M., & Revnivtsev, M. (2001). Soft state of Cygnus X-1: stable disc and unstable corona. *Monthly Notices of the Royal Astronomical Society*, *321*, 759–766.

Cid Fernandes, R., Schlickmann, M., Stasinska, G., Asari, N. V., Gomes, J. M., Schoenell, et al (2009). The starburst-AGN disconnection: LINERs as retired galaxies. In W. Wang, Z. Yang, Z. Luo, & Z. Chen (Eds) The Starburst-AGN Connection, Astronomical Society of the Pacific Conference Series, vol. 408, p. 122

Constantin, A., Green, P., Aldcroft, T., Kim, D.-W., Haggard, D., Barkhouse, W., et al. (2009). Probing the balance of AGN and star-forming activity in the local Universe with ChaMP. *The Astrophysical Journal*, *705*, 1336–1355.

Done, C., & Gierliński, M. (2006). Truncated disc versus extremely broad iron line in XTE J1650–500. *Monthly Notices of the Royal Astronomical Society*, *367*, 659–668.
Emmanoulopoulos, D., Papadakis, I. E., McHardy, I. M., Arévalo, P., Calvelo, D. E., & Uttley, P. (2012). The 'harder when brighter' X-ray behaviour of the low-luminosity active galactic nucleus NGC 7213. *Monthly Notices of the Royal Astronomical Society*, *424*, 1327–1334.
Esin, A. A., McClintock, J. E., & Narayan, R. (1997). Advection-dominated accretion and the spectral States of black hole X-ray binaries: application to Nova MUSCAE 1991. *The Astrophysical Journal*, *489*, 865.
Giacconi, R., Gursky, H., Paolini, F. R., & Rossi, B. B. (1962). Evidence for X-rays from sources outside the solar system. *Physical Review Letters*, *9*, 439–443.
Gierliński, M., & Done, C. (2004). Is the soft excess in active galactic nuclei real? *Monthly Notices of the Royal Astronomical Society*, *349*, L7–L11.
Gierliński, M., & Newton, J. (2006). X-ray spectral transitions of black holes from RXTE All-Sky monitor. *Monthly Notices of the Royal Astronomical Society*, *370*, 837–844.
Gierliński, M., Zdziarski, A. A., Poutanen, J., Coppi, P. S., Ebisawa, K., & Johnson, W. N. (1999). Radiation mechanisms and geometry of Cygnus X-1 in the soft state. *Monthly Notices of the Royal Astronomical Society*, *309*, 496–512.
González-Martín, O., Masegosa, J., Márquez, I., & Guainazzi, M. (2009). Fitting liner nuclei within the active galactic nucleus family: a matter of obscuration? *The Astrophysical Journal*, *704*, 1570–1585.
Gu, M., & Cao, X. (2009). The anticorrelation between the hard X-ray photon index and the Eddington ratio in low-luminosity active galactic nuclei. *Monthly Notices of the Royal Astronomical Society*, *399*, 349–356.
Heckman, T. M. (1980). An optical and radio survey of the nuclei of bright galaxies—activity in normal galactic nuclei. *Astronomy & Astrophysics*, *87*, 152–164.
Ho, L. C., Filippenko, A. V., & Sargent, W. L. W. (1997). A Search for "Dwarf" Seyfert Nuclei. V. Demographics of nuclear activity in nearby Galaxies. *The Astrophysical Journal*, *487*, 568.
Hoyle, F., & Lyttleton, R. A. (1939). The effect of interstellar matter on climatic variation. *Proceedings of the Cambridge Philosophical Society*, *35*, 405.
Ichimaru, S. (1977). Bimodal behavior of accretion disks—theory and application to Cygnus X-1 transitions. *The Astrophysical Journal*, *214*, 840–855.
Iwasawa, K., Fabian, A. C., & Matt, G. (1997). The iron K line complex in NGC1068: implications for X-ray reflection in the nucleus. *Monthly Notices of the Royal Astronomical Society*, *289*, 443–449.
Kotov, O., Churazov, E., & Gilfanov, M. (2001). On the X-ray time-lags in the black hole candidates. *Monthly Notices of the Royal Astronomical Society*, *327*, 799–807.
Lamer, G., McHardy, I. M., Uttley, P., & Jahoda, K. (2003). X-ray spectral variability of the Seyfert galaxy NGC 4051. *Monthly Notices of the Royal Astronomical Society*, *338*, 323–330.
Lyubarskii, Y. E. (1997). Flicker noise in accretion discs. *Monthly Notices of the Royal Astronomical Society*, *292*, 679.
Maccarone, T. J. (2003). Do X-ray binary spectral state transition luminosities vary? *Astronomy & Astrophysics*, *409*, 697–706.
Magdziarz, P., & Zdziarski, A. A. (1995). Angle-dependent Compton reflection of X-rays and gamma-rays. *Monthly Notices of the Royal Astronomical Society*, *273*, 837–848.
Matt, G., Guainazzi, M., Perola, G. C., Fiore, F., Nicastro, F., Cappi, M., et al. (2001). The complex iron line of NGC 5506. *Astronomy & Astrophysics*, *377*, L31–L34.
Mauche, C. W., Liedahl, D. A., Akiyama, S., and Plewa, T. (2008). The Physics of Wind-Fed Accretion. In Axelsson, M., editor, American Institute of Physics Conference Series, volume 1054 of American Institute of Physics Conference Series, pages 3–11.
Narayan, R. (1996). Advection-dominated models of luminous accreting black holes. *The Astrophysical Journal*, *462*, 136.

Narayan, R., Mahadevan, R., and Quataert, E. (1998). Advection-dominated accretion around black holess. In Abramowicz, M. A., Bjornsson, G., and Pringle, J. E., editors, Theory of Black Hole Accretion Disks, page 148.

Narayan, R., & Yi, I. (1994). Advection-dominated accretion: a self-similar solution. *The Astrophysical Journal Letters*, *428*, L13–L16.

Osterbrock, D. E. (1977). Spectrophotometry of seyfert 1 galaxies. *The Astrophysical Journal*, *215*, 733–745.

Paczyński, B. (1971). Evolutionary processes in close binary systems. *Annual Review of Astronomy and Astrophysics*, *9*, 183.

Ponti, G., Miniutti, G., Cappi, M., Maraschi, L., Fabian, A. C., & Iwasawa, K. (2006). XMM-Newton study of the complex and variable spectrum of NGC 4051. *Monthly Notices of the Royal Astronomical Society*, *368*, 903–916.

Qiao, E., & Liu, B. F. (2010). The Spectral features of disk and Corona with mass evaporation in the low/hard state. *Publications of the Astronomical Society of Japan*, *62*, 661.

Schmidt, M. (1963). 3C 273: a star-Like object with large red-shift. *Nature*, *197*, 1040.

Seyfert, C. K. (1943). Nuclear emission in spiral Nebulae. *The Astrophysical Journal*, *97*, 28.

Shakura, N. I., & Sunyaev, R. A. (1973). Black holes in binary systems observational appearance. *Astronomy & Astrophysics*, *24*, 337–355.

Shapiro, S. L., Lightman, A. P., & Eardley, D. M. (1976). A two-temperature accretion disk model for Cygnus X-1—structure and spectrum. *The Astrophysical Journal*, *204*, 187–199.

Shemmer, O., Brandt, W. N., Netzer, H., Maiolino, R., & Kaspi, S. (2006). The hard X-ray spectral slope as an accretion rate indicator in radio-quiet active Galactic nuclei. *The Astrophysical Journal Letters*, *646*, L29–L32.

Sobolewska, M. A., & Papadakis, I. E. (2009). The long-term X-ray spectral variability of AGN. *Monthly Notices of the Royal Astronomical Society*, *399*, 1597–1610.

Sobolewska, M. A., Papadakis, I. E., Done, C., & Malzac, J. (2011). Evidence for a change in the X-ray radiation mechanism in the hard state of Galactic black holes. *Monthly Notices of the Royal Astronomical Society*, *417*, 280–288.

Taniguchi, Y., Shioya, Y., & Murayama, T. (2000). Poststarburst models of LINERS. *Astronom Journal*, *120*, 1265–1272.

Urry, C. M., & Padovani, P. (1995). Unified schemes for radio-loud active Galactic nuclei. *Publications Astronomical Society of the Pacific*, *107*, 803.

Uttley, P., & McHardy, I. M. (2001). The flux-dependent amplitude of broadband noise variability in X-ray binaries and active galaxies. *Monthly Notices of the Royal Astronomical Society*, *323*, L26–L30.

Wu, Q., & Gu, M. (2008). The X-Ray spectral evolution in X-Ray binaries and Its application to constrain the black hole mass of ultraluminous X-ray sources. *The Astrophysical Journal*, *682*, 212–217.

Wu, Y. X., Belloni, T. M., & Stella, L. (2010). Fast spectral variability from Cygnus X-1. *Monthly Notices of the Royal Astronomical Society*, *408*, 2413–2425.

Younes, G., Porquet, D., Sabra, B., & Reeves, J. N. (2011). Study of LINER sources with broad H$\alpha$ emission, X-ray properties and comparison to luminous AGN and X-ray binaries. *Astronomy & Astrophysics*, *530*, A149.

Zdziarski, A. A., Poutanen, J., Paciesas, W. S., & Wen, L. (2002). Understanding the long-term spectral variability of Cygnus X-1 with burst and transient source experiment and all-sky monitor observations. *The Astrophysical Journal*, *578*, 357–373.

# Chapter 2
# AGN: Absorption, Spectral Classification and the Optical Emission Lines

*Scientific theory is a contrived foothold in the chaos of living phenomena.*

Wilhelm Reich (1897–1957)

## 2.1 Synopsis

In this chapter I examine the properties of a sample of nearby galaxies through the use of an archival *Chandra* snapshot survey, with particular emphasis on absorption and the identification of Compton-thick sources. I also discuss the importance of the optical emission lines and, in particular, their use in distinguishing direct AGN emission from starburst emission or reprocessed emission from obscured sources. Although this chapter arises from a separate body of work from that described in later chapters, the results are nevertheless relevant to the selection of AGN sources in Chap. 4. In addition, the distribution of $\Gamma$ with respect to $\dot{m}_{\mathrm{E}}$ within this sample (which is discussed at the beginning of Chap. 4) leads naturally to my examination in Chaps. 3 and 4 of how this relationship evolves within individual sources.

## 2.2 Introduction

Active galactic nuclei (AGN) are widely believed to be powered by accretion onto supermassive black holes (SMBHs), and it is therefore natural to expect that their behaviour should, at least to some extent, mirror that of Galactic X-ray binary systems (XRBs). However, the galactic nuclei in which AGN lie are likely to have very different properties from the local environments of Galactic XRBs, with a much greater density of surrounding matter and therefore an increased risk of contamination from nearby sources.



C. Skipper, *Fast Spectral Variability in the X-ray Emission of Accreting Black Holes*, Springer Theses, DOI 10.1007/978-3-319-09587-5_2

Observationally, AGN are divided into several different types depending upon their properties, and some types are further divided into several sub-types. For example, sources from which broad emission lines have been detected are classified as either type 1.0, 1.2, 1.5, 1.8, 1.9 or 2.0 (Osterbrock 1977), depending upon the strength of the H $\alpha$ and H $\beta$ lines. It has been suggested (e.g. Antonucci and Miller 1985) that these different classes can be unified within a single scenario in which the orientation of the system with respect to the observer is the main factor in producing the different observed properties. The original proposal for a unified scheme was based upon the discovery of polarised broad lines in the optical spectrum of the Seyfert-2 galaxy NGC 1068 and, since then, polarised broad lines have been discovered in $\sim$50 % of Seyfert-2 galaxies for which we have suitable spectropolarimetric data (Tran 2001). However, in recent years a number of sources have been identified that appear to be inconsistent with the unified model (e.g. Pappa et al. 2001; Panessa and Bassani 2002; Georgantopoulos and Zezas 2003; Bianchi et al. 2008) and, in particular, very little is known about the degree to which the unified model applies to low luminosity AGN (LLAGN).

In order to compare the behaviour of AGN to that of XRBs it is therefore crucial that we understand the nature of the objects that we are studying. For instance, it has been suggested that low-ionisation nuclear emission line regions (LINERs) are the AGN equivalent of hard-state Galactic XRBs (e.g. Narayan and McClintock 2008; Constantin et al. 2009; Younes et al. 2011), and should therefore be described using a radiatively inefficient accretion flow (RIAF) model (Narayan et al. 1998; Quataert 2001). Another potential source of confusion when studying AGN is the degree of absorption affecting our view of the source, and recognising the difference between AGN emission that is unabsorbed and emission that is reprocessed (by reflection or scattering, for example) has often proved more difficult than one would imagine.

The majority of nearby galaxies are classified as having H II region nuclei, and it is expected that star formation is likely to play a significant role in most (if not all) of their properties. H II region galaxies are almost always late-type spirals with relatively small black hole masses ($M_{\rm BH} \leq 10^7 M_{\odot}$) and few of these sources contain unresolved nuclear emission. However, bridging the gap between H II region galaxies and AGN are the transition region galaxies; these sources are much more likely to have unresolved nuclei, and possibly represent the "switching on" of AGN in very dense nuclei that are dominated by star formation.

Possibly the greatest question mark surrounding the nature of LLAGN concerns LINERs. Whereas there is little doubt that LINER nuclei that exhibit broad emission lines are powered by AGN (Younes et al. 2011), the origin of the emission in narrow-line LINERs (type 2) is more open to debate. González-Martín et al. (2009b) found that 80 % of their sample of LINERs were AGN, but many alternative theories have been proposed to explain the emission from these objects, such as poststarbursts (Taniguchi et al. 2000) or old stellar populations (Cid Fernandes et al. 2009). González-Martín et al. (2009a) also noted that LINERs may be more heavily obscured than Seyferts. The presence of hard X-ray emission in the nucleus of a LINER is a strong indication of AGN activity, but it does not necessarily follow that the AGN emission is responsible for ionising the narrow line region (NLR). In general, type-2 LINERs are regarded as something of a "mixed bag" and certainly cannot be understood without due consideration of the wider nuclear environments in which they are found.

## 2.3 The Sample

My sample is based upon the well studied Palomar sample of local galaxies, which was the subject of an extensive spectroscopic survey by Ho et al. (1997a). This sample contains 486 northern galaxies ($\delta \geq 0°$), brighter than $B_T = 12.5$ mag and no further in distance than ~100 Mpc, and provides a variety of nuclear spectral types (H II region nuclei, transition region nuclei, LINERs and Seyfert nuclei).

A nearby galaxy sample such as the one I have chosen to use offers a number of advantages when studying AGN activity. Firstly, at distances of less than ~100 Mpc the host galaxies are easily detectable by ground-based surveys such as Palomar or the Sloan Digital Sky Survey (DSS), and, consequently, the sample is complete and does not suffer from the selection effects that plague surveys at larger redshifts. Secondly, nearby galaxies allow us to probe accretion rates that are several orders of magnitude below the Eddington limit and therefore examine accretion on vastly different scales within the same survey. In addition, the angular resolution of *Chandra* allows us to separate emission associated with the SMBH from the many other sources that surround it and therefore reduce the contamination from nearby XRBs and ultra-luminous X-ray sources (ULXs).

In addition to the main sample, I have also included a small number of Palomar Green (PG) quasars (namely PG 0157+001, PG 0844+349, PG 1004+130, PG 1100+772, PG 1115+407, PG 1211+143, PG 1404+226, PG 1512+370, PG 1543+489 and PG 2112+059). Although this chapter focuses heavily upon the Palomar sample, which provides an unbiased view of accretion in the local Universe, the quasars can help provide the high accretion rate perspective when included on the same plots. The distances were estimated using redshift data from the NASA/IPAC Extragalactic Database (NED, with the Hubble constant assumed to be $H_0 = 73\,\mathrm{km\,s^{-1}\,Mpc^{-1}}$), and the black hole masses were taken from Inoue et al. (2007).

## 2.4 The Optical Emission Lines from AGN

The first step towards understanding AGN comes not from X-ray studies but from an appreciation of their optical spectra. Fortunately, much work has already been done in analysing the optical spectra of the Palomar galaxies, and this has been published in a series of papers (Ho et al. 1997a, b, c, 2009). Figure 2.1 shows the optical spectra of a selection of LINER (left) and Seyfert-1 (right) galaxies (with the stellar contribution subtracted), and highlights some of the emission lines that are

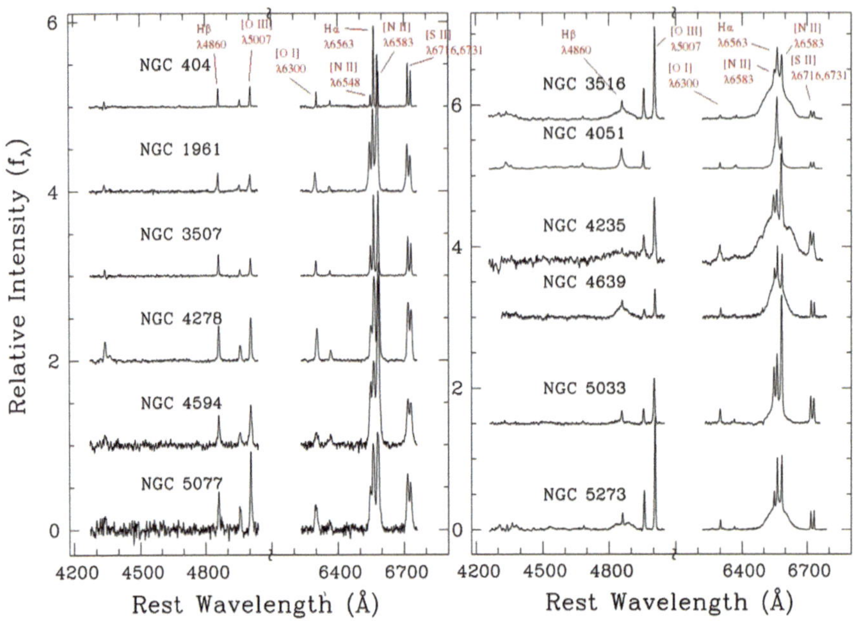


**Fig. 2.1** Optical AGN spectra from Ho et al. (1997a) for a selection of LINERs (*left*) and Seyfert-1s (*right*), offset from each other for clarity. The authors have subtracted the stellar contribution from each source by using various nearby absorption line galaxies as templates. The higher ionisation of Seyferts compared to LINERs is clear from the relative strengths of the [O III] and [O I] lines. Both classes can show broad H $\alpha$ and H $\beta$ lines, but these tend to be stronger in the Seyferts. © AAS. Reproduced with permission

most relevant to this chapter. It can clearly be seen from the relative strengths of the [O III] and [O I] lines that LINERs exhibit lower ionisation that Seyferts, but it is also clear that LINERs can show weak broad H $\alpha$ lines. The presence of broad lines is probably the most reliable indicator that the emission is from the vicinity of a SMBH, and in Seyfert galaxies the broadening can be as much as FWHM $\approx$ 7,600 km s$^{-1}$ (Ho et al. 1997c).

## 2.5 Classifying Low Luminosity AGN

Many galaxies in the nearby Universe show emission lines from their nuclei (Ho et al. 1997a detected emission lines in 421 of their 486 sample galaxies), and accretion onto a SMBH is not the only possible explanation for this phenomenon. For instance, the ionisation could be the result of nuclear starbursts, in which a large number of stars are born in a compact area in a short period of time, and subsequently ionise the surrounding gas with powerful winds. Alternatively, it could be the result of a

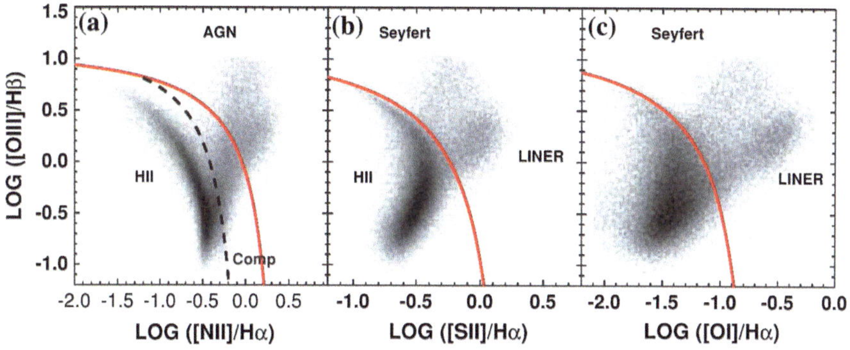


**Fig. 2.2** The relative strengths of the emissions lines used as a diagnostic between ionisation due to star formation and accretion (Kewley et al. 2006). Reproduced with permission of Oxford University Press on behalf of the Royal Astronomical Society

high density of XRBs or supernova remnants in the region surrounding the galactic nuclei.

To help make sense of the variety of different properties observed amongst LLAGN, these sources are usually classified using a plot similar to those proposed by Baldwin et al. (1981). This kind of diagnostic diagram has become known as a BPT diagram, and these plots utilise various different emission lines in order to differentiate between ionisation due to star formation and accretion. Typical diagnostics include [N II]/H $\alpha$, [S II]/H $\alpha$ or [O I]/H $\alpha$ versus [O III]/$H\ \beta$, as shown in Fig. 2.2 (Kewley et al. 2006).

The Palomar data used within this chapter have already been classified into their various spectral types, and these are supplied in the same paper as the emission line strengths (Ho et al. 1997a). However, these classifications are based upon line strength ratios that have not been corrected for reddening (see Richings et al. 2011) and, therefore, I instead choose to use classifications that are based upon the extinction-corrected data. These data have been provided by Phil Uttley (private communication), along with the de-reddened emission line strengths. The latter were corrected for extinction using the extinction law of Cardelli et al. (1989), and spectral classifications assigned to each source using the BPT diagnostic diagrams of Kewley et al. (2006). Figure 2.3 shows the [N II]/H $\alpha$ versus [O III]/H $\beta$ diagnostic for the complete Palomar sample.

Using this new classification scheme based upon extinction-correction line ratios I find that I have 142 H II region nuclei, 73 transition region nuclei, 134 LINERs, 56 Seyfert nuclei and 81 unknowns. The 'unknowns' include the 79 galaxies from which Ho et al. (1997a) detected no visible emission lines (or only very weak emission lines).

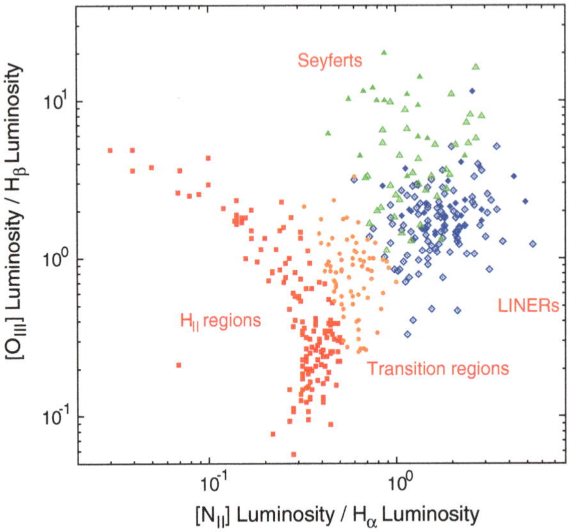


**Fig. 2.3** Re-classified, extinction corrected BPT diagram for the complete Palomar sample. *Red squares* are H II regions, *orange circles* are transition regions, *blue diamonds* are LINERs (*filled* are type 1, *unfilled* type 2) and *green triangles* are Seyferts (*filled* are type 1, *unfilled* type 2). The H II regions sweep from the *top left* to the *bottom* of the plot (referred to as the star-forming sequence by Kewley et al. (2006)) and the Seyferts and LINERs occupy the *upper-right* region of the plot

## 2.6 The Optical Lines as Indicators of X-ray Luminosity

The X-ray luminosity from AGN, and especially at energies above 2 keV, is often used as a reliable indicator of accretion rate. The reason that this region of the electromagnetic spectrum is particularly suitable for this purpose is due to the origin of the X-ray emission, which is believed to be produced on scales comparable to the accretion flow through processes such as inverse Compton scattering of lower energy seed photons in a corona of hot electrons or blackbody emission from the accretion disc. Furthermore, the energetic nature of X-ray photons enables them to penetrate the often dense clouds of gas and dust that may surround a SMBH and therefore provide a direct view of the innermost regions of the system.

In some sources, however, the column density ($N_{\mathrm{H}}$) is so high (presumably the result of matter predominantly in the vicinity of the SMBH) that not even X-ray photons in the typical 0.3–10 keV sensitivity range of a modern X-ray telescope such as *Chandra* are able to penetrate. In this scenario, which occurs when $N_{\mathrm{H}}$ exceeds $\sim 10^{24}\,\mathrm{cm}^{-2}$, the source is said to be Compton thick and the intrinsic X-ray flux below 10 keV is completely absorbed, replaced instead by a combination of scattered and reflected emission which is expected to be a couple of orders of magnitude weaker than the intrinsic emission. However, under the standard model of AGN (Antonucci and Miller 1985) the absorbing matter takes the form of a torus, and therefore a column density $\geq 10^{24}\,\mathrm{cm}^{-2}$ in the observer's line of sight will not

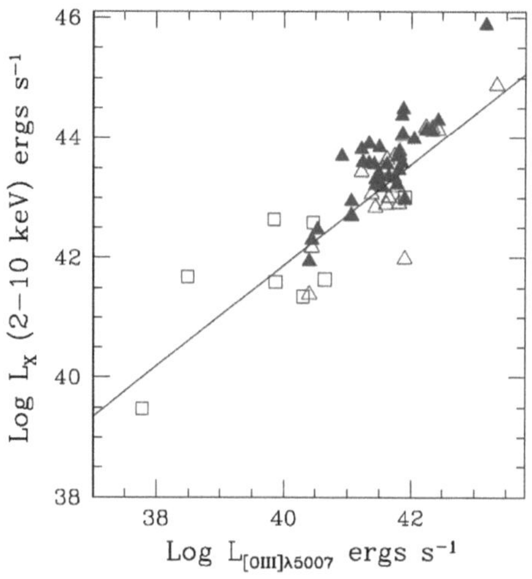


**Fig. 2.4** Plot of $L_{[\mathrm{O\,III}]}$ versus $L_X$ for three samples of Seyfert-1 (Georgantopoulos and Akylas 2010). Reproduced with permission © ESO

prevent the ionisation of the NLR which surrounds the whole system to a distance of up to ~100 pc.

The strengths of various optical emission lines originating in the NLR, in particular the Balmer lines (H $\alpha$ and H $\beta$) and [O III]$_{\lambda5007}$, have often been used as good isotropic indicators of X-ray luminosity or AGN power (e.g. Panessa et al. 2006; González-Martín et al. 2009b). The strength of the correlation between the 2–10 keV X-ray luminosity and the optical lines is particularly strong amongst sources that also exhibit broad emission lines in their optical spectra, the presence of which is probably the clearest indication one can get that an AGN is responsible for the X-ray emission. Figure 2.4 shows the correlation between $L_{2-10\,\mathrm{keV}}$ and $L_{[\mathrm{O\,III}]\lambda5007}$ for three samples of Seyfert-1, demonstrating the tight correlation that is often found between these properties in unobscured sources. Many authors have also examined the correlation between X-ray luminosity and H $\alpha$ luminosity (e.g. Ho et al. 2001; Panessa et al. 2006; Terashima et al. 2000) and found that it persists down to bolometric luminosities as low as $10^{37}$ erg s$^{-1}$.

A reliable indicator of intrinsic X-ray luminosity is a valuable tool in identifying Compton-thick AGN. The standard model predicts that the NLRs of sources that are heavily obscured ($N_{\mathrm{H}} \geq 10^{24}$ cm$^{-2}$) by the torus will still be ionised by the X-ray emission from the central engine along the less obscured axis perpendicular to the observer's line of sight. These Compton-thick sources will be expected to show $L_{[\mathrm{O\,III}]}/L_X$ ratios that are significantly greater than for those where lower obscuration allows the intrinsic X-ray emission to be viewed directly.

## 2.7 Data Reduction

### 2.7.1 Observation Selection

In order to ensure a consistent approach throughout it was decided to use only observations made with the Advanced CCD Imaging Spectrometer (ACIS) instrument and neglect those that used the High Resolution Camera (HRC), which, in any case, has rarely been used to observe any of the sources in my sample. The *Chandra* archive was searched for each of the sources and suitable observations were identified and retrieved. In general, observations with particularly short exposure times ($<\sim 2$ ks) were ignored, but, in anticipation of the effect of pileup on the brighter sources, observations with short frame times or those which utilised the High Energy Transmission Grating (HETG) were also desirable.

### 2.7.2 Image Analysis

For each observation the level 2 event data were examined in DS9 and source and background regions were placed around the X-ray point source that was thought to represent emission from the central engine. In all cases I took the presence of an unresolved nuclear source in the 2–10 keV band as evidence of AGN activity, and, although identification of the nucleus was generally straightforward, for some sources it was necessary to reference the literature for a more precise location. In most cases a circle of radius $1.5''$ was used to describe the source region and an annulus of between 8 and $12''$ was used for the background region, but this varied slightly depending upon the proximity of nearby sources. Observations in which no unresolved emission was detected were ignored, as were those of NGC 4151 and NGC 4486 due to the severe effects of pileup. An unresolved X-ray nucleus could be identified with confidence in 89 sources.

Data reduction was performed using CXC *Chandra* Interactive Analysis of Observations (CIAO) software version 4.2. After restricting the event files to the energy range 0.3–10 keV the script PSEXTRACT was then used to automatically generate grouped spectra with 15 counts per channel. In most cases it was necessary to recreate the response matrix file (RMF) and ancillary response function (ARF) files manually using the CIAO tools MKACISRMF and MKARF respectively.

### 2.7.3 Spectral Fitting

Once spectra had been extracted the fitting was performed using XSPEC version 12.6.0 over the energy range 0.3–10 keV. Each source was initially fitted with a model comprising Galactic absorption (PHABS) fixed to column densities obtained using the NH program from the NASA HEASARC FTOOLS software[1] (Blackburn 1995), a power law component (ZPOWERLW) and a second absorption component (ZPHABS) representing absorption in the host galaxy. In some cases additional components were included such as an iron K $\alpha$ line at 6.4 keV (modelled with ZGAUSS) or a thermal component (MEKAL) but these were only used if the spectrum had sufficient counts to justify the extra parameters and the quality of the fit was significantly improved.

For each observation the ratio of the bolometric luminosity to the Eddington luminosity serves as a proxy for accretion rate (see Chap. 1 for a discussion of how luminosity relates to $\dot{m}_{\rm E}$), and is estimated using

$$\frac{L_{\rm bol}}{L_{\rm Edd}} = \frac{F_{2-10\,{\rm keV}} \times 4\pi d^2}{L_{\rm Edd}} \times C_{2-10\,{\rm keV}} \tag{2.1}$$

where $F_{2-10\,{\rm keV}}$ is the 2–10 keV flux from the *Chandra* fit, $d$ is the distance to the galaxy (from Ho et al. 1997a), $C_{2-10\,{\rm keV}}$ is the bolometric correction (see below) and $L_{\rm Edd}$ is the Eddington limit for the source given by

$$L_{\rm Edd} = \frac{4\pi\, G M_{\rm BH}\, m_{\rm p}\, c}{\sigma_{\rm T}} \tag{2.2}$$

where $G$ is the gravitational constant, $M_{\rm BH}$ is the black hole mass, $m_{\rm p}$ is the mass of a proton and $\sigma_{\rm T}$ is the Thompson scattering cross-section for an electron. For each source, $M_{\rm BH}$ was estimated using the relation between black hole mass and velocity dispersion ($\sigma$, obtained for the Palomar sample from Ho et al. (2009)) given by Tremaine et al. (2002), namely

$$\log\frac{M_{\rm BH}}{M_\odot} = \alpha + \beta \log\frac{\sigma}{\sigma_0} \tag{2.3}$$

where $\alpha = 8.13 \pm 0.06$, $\beta = 4.02 \pm 0.32$ and $\sigma_0 = 200\,{\rm km\,s^{-1}}$. The bolometric correction, $C_{2-10\,{\rm keV}}$, is estimated using the relation from Marconi et al. (2004), rearranged to

$$\log C_{2-10\,{\rm keV}} = \log\frac{L_{\rm bol}}{L_{2-10\,{\rm keV}}}$$

$$\log C_{2-10\,{\rm keV}} = 2.98 - 0.264\log L_{\rm bol} + 0.03\log^2 L_{\rm bol} - 0.0015\log^3 L_{\rm bol} \tag{2.4}$$

where $L_{\rm bol}$ is the bolometric luminosity in units of $L_\odot$. Equation 2.4 must be solved iteratively in order to find $L_{\rm bol}$ from the known $L_{2-10\,{\rm keV}}$.

[1] http://heasarc.gsfc.nasa.gov/ftools/.

## 2.8 Distribution of $M_{\rm BH}$, Eddington Ratio and $N_{\rm H}$ for the Palomar Sample

Amongst the properties derived directly from the spectral fits to the *Chandra* data are the source flux ($F_{2-10\,\rm keV}$), the photon index ($\Gamma$), the iron K $\alpha$ line width and the absorbing column density ($N_{\rm H}$). To these results I add data from the literature, such as the colour excess ($E(B-V)$), host galaxy morphology, emission line luminosity (H $\alpha$, H $\beta$, [O III]$_{\lambda 5007}$, [N II]$_{\lambda 6583}$ and the sum of [S II]$_{\lambda 6716}$ and [S II]$_{\lambda 6731}$), source distance (all Ho et al. 1997a), velocity dispersion ($\sigma$) (Ho et al. 2009) and 15 GHz radio luminosity (Nagar et al. 2005). From these data I then derive the luminosity ($L_{2-10\,\rm keV}$), the black hole mass ($M_{\rm BH}$) and the Eddington ratio ($\lambda = L_{\rm bol}/L_{\rm Edd}$).

Figure 2.5 (left) shows a histogram of $M_{\rm BH}$ for the complete Palomar sample, broken down by spectral classification. The range of masses over which each spectral type is distributed can be explained by the possible evolutionary sequence from H II regions to transition regions, to Seyfert galaxies, and finally to LINERs (e.g. Constantin et al. 2009). In particular, there is very little overlap in $M_{\rm BH}$ between the LINERs and H II regions, both of which occupy the low accretion rate end of the scale ($\lambda \leq 10^{-3}$).

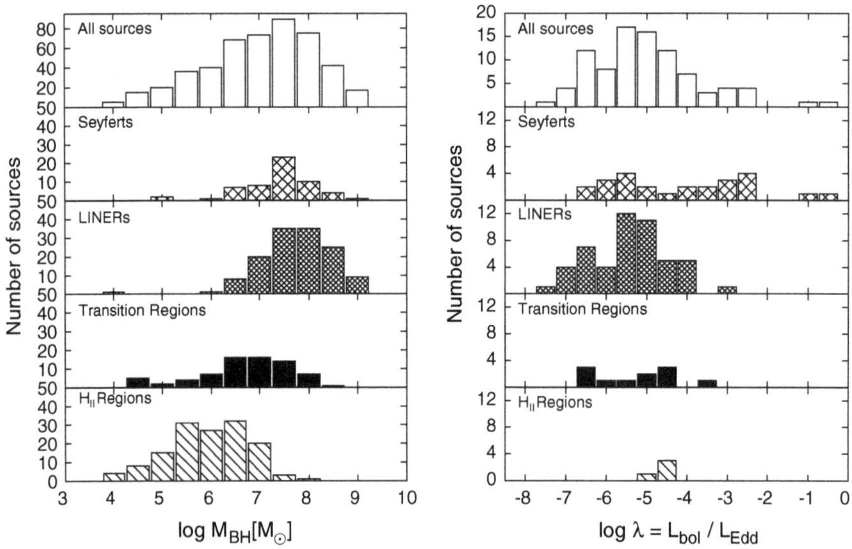


**Fig. 2.5** *Left* Histogram of black hole mass for the complete sample. LINERs and H II regions generally do not overlap much in black hole mass, with the latter occupying the high-mass end of the scale. *Right* Histogram of Eddington ratio from the *Chandra* analysis of the sample. The Seyferts extend over a much larger range of $\lambda$ than the other spectral types, and appear to show a slightly bimodal distribution (the distribution can be fitted with two Gaussian components, centred on $\log \lambda = -2.95$ and $-5.65$; I find that an F-test rejects a single-Gaussian fit at the 1-$\sigma$ level); both of these facts can be attributed to absorption, with the unobscured Seyfert-1s mostly occupying the $\lambda \geq 10^{-4}$ end of the scale and the Seyfert-2s the $\lambda \leq 10^{-4}$ end

Figure 2.5 (right) shows a histogram of Eddington ratio ($\lambda = L_{bol}/L_{Edd}$) from the *Chandra* analysis of the sample. Whereas the transition regions, LINERs and H II regions are generally low Eddington ratio ($\lambda \leq 10^{-3}$), the Seyfert galaxies cover a much wider range spanning several orders of magnitude. The Seyfert galaxies also show a bi-modal distribution in $\lambda$, with the type-1 Seyferts found at high Eddington ratios ($\lambda \geq 10^{-4}$) and the type-2 Seyferts found at low Eddington ratios ($\lambda \leq 10^{-4}$).

The slight differences in $M_{BH}$ and $\lambda$ between different spectral types are more clearly seen when the two parameters are plotted against each other (Fig. 2.6). In this plot I have marked regions where the different spectral types are generally found, and there appears to be a fairly clear distinction between the quasars, type-1 Seyferts, H II regions and LINERs, with the transition regions and type-2 Seyferts more scattered.

The standard model of AGN asserts that many of the differences in fundamental properties that we observe in these sources can be explained as an effect of inclination, and the lack of broad optical emission lines in type-2 sources is simply a result of obscuration by the torus. If this is true, it would therefore be reasonable to expect that type-2 sources should, on average, exhibit greater absorption than type-1

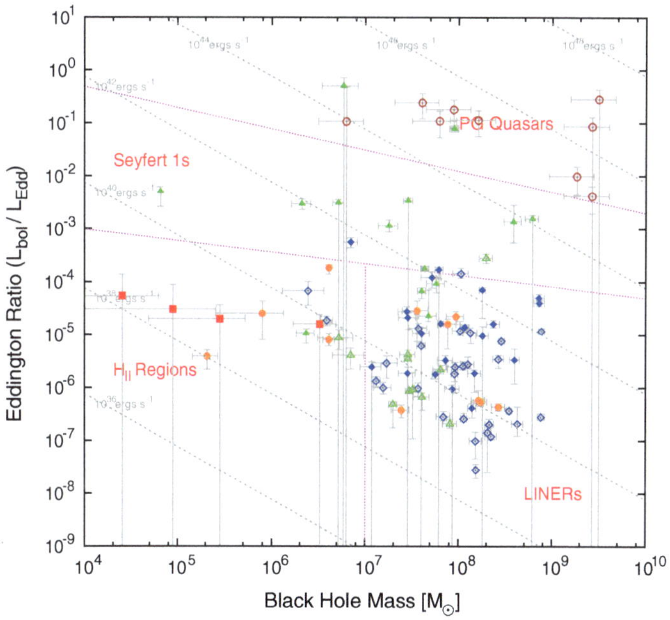


**Fig. 2.6** Plot of black hole mass $M_{BH}$ versus Eddington ratio $\lambda = L_{bol}/L_{Edd}$ (serving as a proxy for accretion rate) for the Palomar sample galaxies and a small selection of PG quasars. The quasars (*open red circles*), Seyfert-1s (*filled green triangles*), LINERs (*filled* and *open blue circles*, representing type 1 and type 2 respectively) and H II region nuclei (*red squares*) are generally found in different regions of the plot. The remaining symbols represent Seyfert-2s (*open green triangles*) and transition region nuclei (*orange circles*)

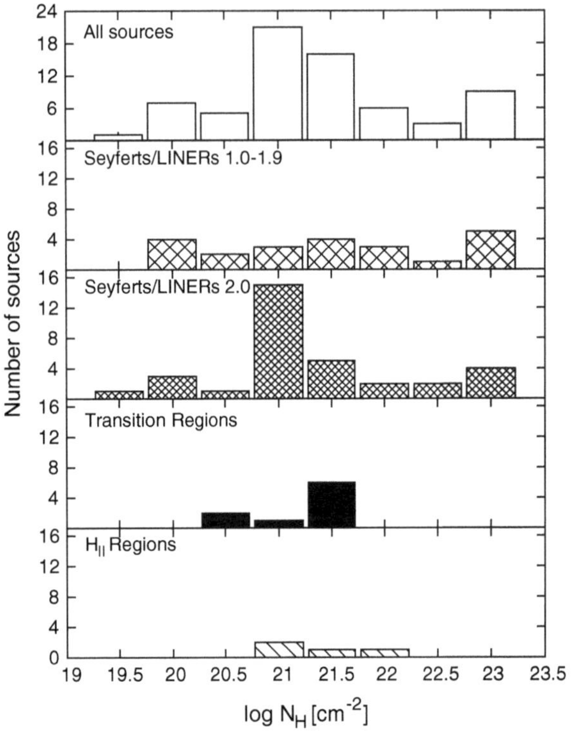


**Fig. 2.7** Histogram of X-ray column density ($N_H$) for the complete sample, and also broken down by spectral type. Seven sources were excluded because the fits to the soft emission were too uncertain to obtain a reliable estimate of the absorption, and a further ten sources (7 type-1, 2 type-2 and 1 transition region) appeared to show no absorption

sources. However, there is always a degree of uncertainty over any measurement of the column density, especially when $N_H \leq \sim 10^{23}$ cm$^{-2}$; to fit an absorption model one must know the exact form of the underlying continuum, a challenge that is made more difficult by the presence of soft emission of a fairly uncertain nature in many AGN.

Figure 2.7 shows a histogram of the absorbing column density ($N_H$), measured directly from the *Chandra* spectral fits. The type-2 AGN, transition regions and H II regions generally show a quite narrow range of absorption, with 30 of the 46 sources (65 %) from which the absorption could be measured having a column density of between $10^{20.75}$ and $10^{21.75}$ cm$^{-2}$. This level of absorption is within an order of magnitude of Galactic absorption, and is therefore not unexpected in the region of a galaxy nucleus; a column density of $\sim 10^{21}$ cm$^{-2}$ could easily be produced by dust lanes, for instance.

In contrast, the type-1 sources exhibit a much greater variety of column densities and show no evidence of being less absorbed than their type-2 counterparts.

Figure 2.7, however, does not show the intrinsic absorption of sources that are Compton thick; these sources may instead appear on the plot with much lower column densities, representing the absorption that affects the reprocessed emission as X-rays are scattered or reflected into our line of sight. Furthermore, there are ten sources (7 type-1, 2 type-2 and 1 transition region) where the absorption is too low to be detectable (usually due to having too few photons to avoid heavy rebinning of the data), and these have been excluded from the plot.

## 2.9 Identifying the Compton-Thick AGN in Our Sample

It is unfortunate that, in order to assess the effectiveness of a particular technique for identifying Compton-thick AGN, one must first apply another—possibly equally uncertain—technique in order to determine which members of the sample are indeed Compton thick. A good indicator of heavy absorption is the presence of polarised broad lines in the optical spectrum (Antonucci and Miller 1985), and when combined with an X-ray column density that appears very low then Compton thickness is almost assured. However, the detection of polarised broad lines is only possible in the brightest sources and there is currently very little spectropolarimetric data in the literature for sources within my sample.

Another useful indicator of Compton thickness is the presence of hard X-ray emission ($>10$ keV) from a source which is relatively weak at softer energies; this hard emission can penetrate an absorbing column that is dense enough to block the 2–10 keV emission that is within the peak sensitivity range of observatories such as *Chandra* and *XMM Newton*. Surveys such as the *Swift* BAT survey (Baumgartner et al. 2012) have produced catalogues of X-ray sources at energies above 14 keV, and 17 of the 89 sources for which I have obtained Chandra spectra are found amongst the 1,171 sources detected by BAT. Figure 2.8 shows a plot of the 2–10 keV *Chandra* luminosity from my survey against the 14–195 keV *Swift* BAT luminosity (Baumgartner et al. 2012).

Ultimately, AGN are classified as Compton thick or Compton thin depending upon evidence from multiple indicators and, initially, I choose to select the Compton-thick sources in my sample using the 0.3–10 keV spectra. Figure A.1 in Appendix A shows the *Chandra* spectra of 11 sources that have at least one of two Compton-thick indicators, which are a flat spectrum above 2 keV and a large iron K $\alpha$ line equivalent width. Five of these sources (NGC 1052, NGC 1068, NGC 3079, NGC 4102 and NGC 7479) were also detected by the BAT survey with larger than average ratios of $L_{14-195\,\mathrm{keV}}/L_{2-10\,\mathrm{keV}}$ (Fig. 2.8).

After Compton-thick candidates had been identified from their *Chandra* spectra, I next performed a literature search in order to verify or reject those sources. Aside from the archetypal Compton-thick Seyfert-2, NGC 1068 (for which this step wasn't considered necessary), this provided support for the inclusion of NGC 2782 (Zhang et al. 2006), NGC 3079 (Iyomoto et al. 2001), NGC 3690 (Della Ceca et al. 2008), NGC 4036 (González-Martín et al. 2009a), NGC 4102 (González-Martín et al. 2011),

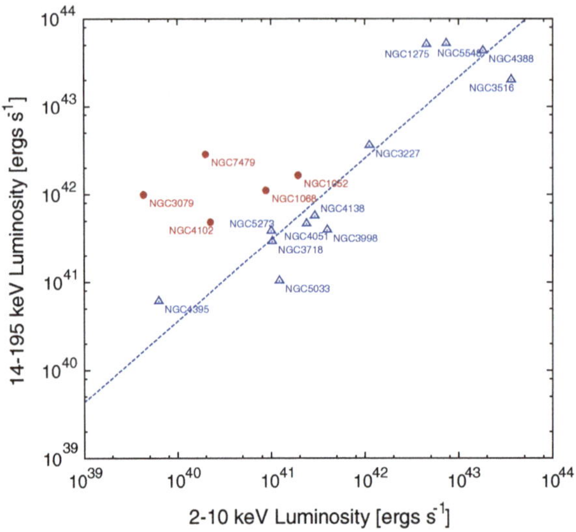


**Fig. 2.8** Plot of the 2–10 keV *Chandra* luminosities from my survey against the 14–195 keV *Swift* BAT luminosities from Baumgartner et al. 2012. Sources with *Chandra* spectra that show signs of Compton thickness, such as a flat spectrum above 2 keV or a large iron K $\alpha$ line equivalent width, are plotted with *filled red circles* and sources that show no such signs are plotted with *open blue triangles*. A linear fit to the latter has equation $\log L_{14-195\,\mathrm{keV}} = 3.54 + 0.93 \log L_{2-10\,\mathrm{keV}}$

NGC 4501 (Brightman and Nandra 2008), NGC 5194 (Cappi 2006) and NGC 7479 (Georgantopoulos et al. 2011). However, the classification of two further sources, NGC 1052 and NGC 4111, is more uncertain. González-Martín et al. (2009a) found that, for both these sources, the only indication of Compton thickness is their unusually flat spectra, and other common indicators suggest that they are in fact Compton thin. In this chapter I include both these sources in my list of likely Compton-thick candidates, but highlight instances where their properties are inconsistent with the other 9 sources.

## 2.10 The Relationship between $L_{2-10\,\mathrm{keV}}$ and $L_{[\mathrm{O\,III}]}$ and its use as an Absorption Indicator

The emission line most commonly used as an indicator of intrinsic X-ray luminosity is the [O III] forbidden line of doubly ionised oxygen (e.g. see Bassani et al. 1999; Georgantopoulos and Akylas 2010; González-Martín et al. 2009b; Panessa and Bassani 2002; Panessa et al. 2006), emitted in the green part of the optical spectrum at 500.7 and 495.9 nm. In this thesis I use only the 500.7 nm line intensities, which were published for the entire Palomar sample by Ho et al. (1997a).

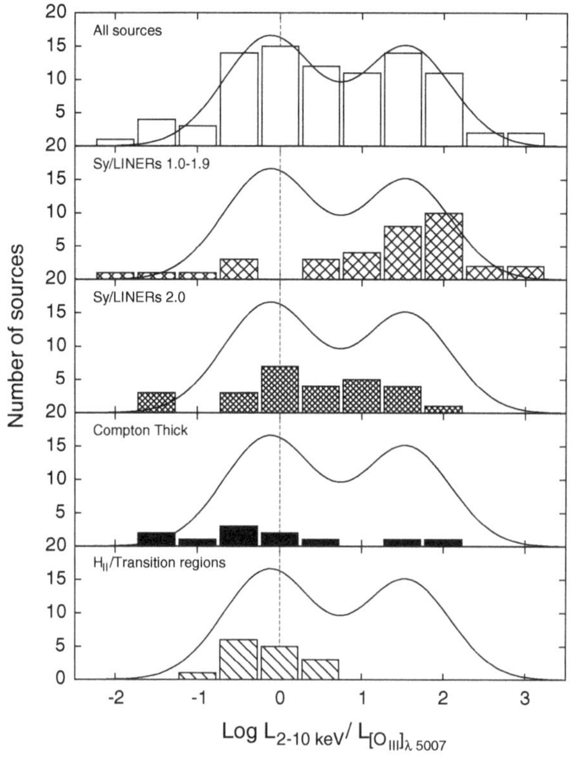


**Fig. 2.9** Histogram showing the ratio of X-ray luminosity to $[\mathrm{O\,III}]_{\lambda 5007}$ luminosity. *Top panel* All sources for which *Chandra* spectra were extracted. The distribution appears slightly bimodal and has been fitted with two Gaussians (centred at −0.129 and 1.549), which are shown in each of the panels (*black line*). *Second panel* Sources classified as either Seyfert 1.0–1.9 or LINER 1.0–1.9. I find a mean $\log_{10} L_{2-10\,\mathrm{keV}}/L_{[\mathrm{O\,III}]}$ ratio of 1.23. *Third panel* Sources classified as either Seyfert-2 or LINER-2. I find a mean $\log_{10} L_{2-10\,\mathrm{keV}}/L_{[\mathrm{O\,III}]}$ ratio of 0.344. *Fourth panel* Sources with X-ray spectra showing evidence of Compton thickness, such as a flat spectrum above 2 keV or a large iron K $\alpha$ line equivalent width. I find a mean $\log_{10} L_{2-10\,\mathrm{keV}}/L_{[\mathrm{O\,III}]}$ ratio of −0.178. *Bottom panel* Sources classified as either H II region or transition region nuclei, in which emission associated with stellar processes would be expected to contribute greatly to the optical emission line strength. I find a mean $\log_{10} L_{2-10\,\mathrm{keV}}/L_{[\mathrm{O\,III}]}$ ratio of −0.199. Compton-thin sources that are powered by an AGN tend to have X-ray/[O III] ratios that are scattered about the second Gaussian peak, whereas sources that are believed to be Compton thick are more scattered. However, the spectral classifications most strongly clustered about the first Gaussian peak are the H II/transition region nuclei, raising the possibility that star formation may be a significant factor in producing the double-Gaussian distribution of the $L_{2-10\,\mathrm{keV}}/L_{[\mathrm{O\,III}]}$ histogram in LLAGN

Previous studies of the relationship between $L_{2-10\,\mathrm{keV}}$ and $L_{[\mathrm{O\,III}]}$ have reported upon a bimodal distribution in the histogram of $R_{\mathrm{X}/[\mathrm{O\,III}]} = L_{2-10\,\mathrm{keV}}/L_{[\mathrm{O\,III}]}$ (e.g. Maiolino et al. 1998; González-Martín et al. 2009b), which the authors associate with the presence of Compton-thick and Compton-thin populations in the sample. In Fig. 2.9 I find only a slight suggestion of bimodality in my data, and fit the histogram of all sources with two Gaussian components (centred at $\log R_{X/[O\,III]} = -0.129$ and 1.549). However, an F-test fails to reject a single-Gaussian fit at the 1–$\sigma$ level.

When the data are broken down by spectral classification it can be seen that the sources for which we can say with greatest certainty that the X-ray emission originates from an accretion flow (Seyferts and LINERs, types 1.0–1.9) show the highest $L_{2-10\,keV}$ to $L_{[O\,III]}$ ratios ($\overline{\log R_{X/[O\,III]}} = 1.23$), and are clustered around the second Gaussian peak. The lowest ratios are found in the H II regions/transition regions and the Compton-thick sources ($\overline{\log R_{X/[O\,III]}} = -0.199$ and $-0.178$ respectively) and the former are strongly clustered about the first Gaussian peak (more so than the Compton-thick sources, which are more spread out). These results suggest that star formation (which is expected to contribute strongly to the [O III] emission of H II region and transition region nuclei) may be responsible for the double-Gaussian distribution of the $L_{2-10\,keV}/L_{[O\,III]}$ histogram in LLAGN, bringing into question the validity of using this ratio as a Compton-thick diagnostic.

In Fig. 2.10 I plot $L_{2-10\,keV}$ against $L_{[O\,III]\lambda5007}$, dividing the sources into type-1 and 2 AGN, starburst sources (H II regions and transition regions) and Compton-thick AGN. The solid red line corresponds to the local minimum between the two

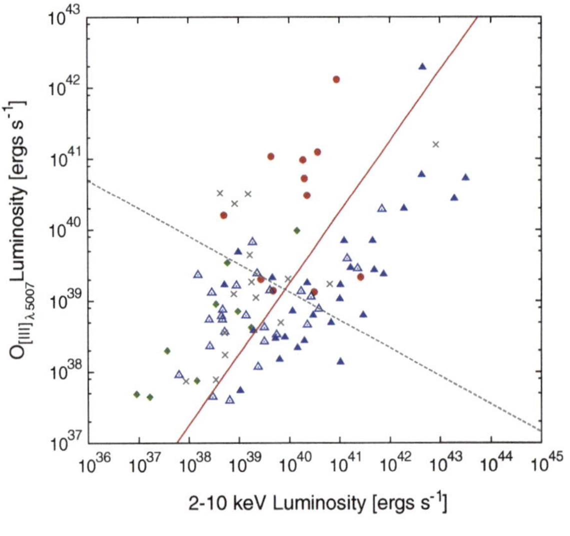


**Fig. 2.10** 2–10 keV X-ray luminosities from my *Chandra* survey plotted against the luminosity of the [O III]$\lambda$5007 emission line (from Ho et al. 1997a). The *red line* ($L_{[O\,III]} = 0.181\,L_{2-10\,keV}$) corresponds to the minimum in the double-Gaussian fit to the $L_{2-10\,keV}/L_{[O\,III]}$ histogram. Type-1.0–1.9 LINERs and Seyferts are shown with *filled blue triangle*, type-2 LINERs and Seyferts with *open blue triangles* and H II/transition region nuclei with *filled green diamonds*. The sources selected as Compton-thick candidates are marked with *filled red circles*. Any observations with fewer than 100 counts, in which the spectra were not of sufficient quality to identify Compton-thick candidates, are marked with *grey crosses*. Most of the Compton-thick sources are found in the region $L_{[O\,III]} \geq 0.181\,L_{2-10\,keV}$, but are clearly distinct in their [O III] luminosity from Compton-thin sources (shown separated by the *dashed grey line*), suggesting that the usefulness of the X-ray/[O III] ratio as a Compton-thick diagnostic may not extend down to LLAGN

Gaussian components that were fitted to the $L_{2-10\,\mathrm{keV}}/L_{[\mathrm{O\,III}]}$ histogram (Fig. 2.9), and most of the Compton-thick candidates lie to the left-hand-side of this line. However, amongst the sources on the left-hand-side of the red line (sources where $L_{[\mathrm{O\,III}]} \geq 0.181\,L_{2-10\,\mathrm{keV}}$), there is a clear distinction in luminosity between Compton-thick and Compton-thin sources, with the former being found at higher luminosities. This division in luminosity between Compton-thick and Compton-thin sources (shown with a dashed grey line) also marks a change in the distribution of sources for the sample as a whole. At higher luminosities there is a noticeable gap in the density of points with mostly type-1 AGN to the right-hand-side of the red line and mostly Compton-thick AGN to the left-hand-side. A histogram of $L_X/L_{[\mathrm{O\,III}]\lambda5007}$ for these data would produce a far clearer bimodal distribution than the one I find in Fig. 2.9. However, at lower luminosities (beneath the dashed grey line) this gap disappears and the data points become more evenly spread. Here I find an absence of Compton-thick sources, replaced instead by a population of starburst sources (H II/transition region nuclei), all of which are found to the left-hand-side of the red line. This result is consistent with recent studies (e.g. Hopkins et al. 2009; Panessa and Bassani 2002) which report that LLAGN that are heavily obscured by the torus are less common than previously thought.

The picture that emerges from these data is that the $L_{2-10\,\mathrm{keV}}/L_{[\mathrm{O\,III}]}$ ratio diagnostic of Compton thickness is only valid above $L_{[\mathrm{O\,III}]} \approx 10^{39}\,\mathrm{erg\,s^{-1}}$. At lower luminosities, starburst contamination creates considerably more scatter in the data and the likely disappearance of the torus means that we no longer see a separate population of sources representing the Compton-thick candidates. In this luminosity regime, the standard model of Antonucci and Miller (1985) no longer applies and the $L_{2-10\,\mathrm{keV}}/L_{[\mathrm{O\,III}]}$ ratio may instead be useful in separating starburst nuclei from AGN.

The similarity of the X-ray to [O III] ratios in Compton-thick sources and starburst sources (H II regions/transition regions) may suggest a connection between the two, further evidence for which is provided by the breakdown of Compton-thick sources by spectral type, which reveals that 33 % (3 from 9) of transition regions for which I have spectra available and at least 100 spectral counts are actually Compton thick. This figure is comparable to that found for Seyferts of type 1.8–2.0, where 23 % (3 from 13) appear Compton thick, and higher than for the other spectral types, with 0 % (0 from 7) of Seyferts of type 1.0–1.5 and 10 % (4 from 41) of LINERs also appearing Compton thick. Only three H II regions have at least 100 spectral counts, and none of these appear Compton thick.

## 2.11 A Comparison of Absorption in the X-ray and Optical Bands

The X-ray emission from an accreting black hole is believed to arise in the innermost regions of the accretion flow ($\leq$1 pc) where the in-falling material is at its hottest. In contrast, the narrow emission lines (such as H$\alpha$, H$\beta$, [O III], [N II], [S II], etc.) observed from AGN are produced in the NLR at distances of up to ~100 pc from the black hole. Therefore, it should be expected that, even in sources where the alignment of the torus is such that it does not obscure the emission from the central engine, the X-ray emitting region should be more heavily obscured than the NLR.

Absorption in the X-ray band is mostly due to the presence of gas between the source and the observer, which is quantified in terms of its hydrogen column density ($N_{\rm H}$) in units of atoms per cm$^2$. Extinction in the optical band is mostly due to dust and is estimated from the colour excess

$$E(B-V) = (B-V)_{\rm observed} - (B-V)_{\rm intrinsic} \tag{2.5}$$

where $B-V$ is the difference between the magnitudes in the $B$ and $V$ filters. The gas and dust that are responsible for X-ray and optical extinction respectively usually exist together in a ratio that is reasonably consistent (at least, the estimates of this ratio from the literature vary by a factor of less than ~5, which is small when compared to the scatter found in my data). In this chapter I choose to use the empirical relationship of Bohlin et al. (1978) ($N_{\rm H}/E(B-V) = 5.8 \times 10^{21}$ cm$^{-2}$ mag$^{-1}$), but I could also have chosen Predehl and Schmitt (1995) ($N_{\rm H}/E(B-V) = 1.79 \times 10^{21}$ cm$^{-2}$ mag$^{-1}$).

Figure 2.11 shows a histogram of the ratio between $N_{\rm H}$ and $E(B-V)$, broken down to separately show Compton-thin sources with low and high X-ray to [O III] ratios (where $L_{\rm [O\,III]} = 0.181\, L_{2-10\,\rm keV}$ is the boundary between the two). Also shown on the same plot is the Bohlin et al. (1978) relationship between $N_{\rm H}$ and $E(B-V)$ (dotted red line). The Seyferts and LINERs with high X-ray to [O III] ratios (i.e. the sources clustered under the second Gaussian peak of Fig. 2.9) have much higher $N_{\rm H}/E(B-V)$ ratios than we would expect if the same material were responsible for both the X-ray and optical extinction. These sources undoubtedly contain a high proportion of AGN, with broad optical lines detected in 23 of 36 sources (64 %) compared to just 5 of 28 (18 %) in the sources with low X-ray to [O III] ratios, and this 'excess' of X-ray absorption is due to the origin of the X-rays being deep within the AGN where the photons are subjected to much greater obscuration than that which affects the narrow line region further out.

## 2.12 The Unusual Case of the Seyfert-2 Galaxy NGC 3147

Figure 2.11 also reveals that there are four sources (NGC 1068, NGC 2782, NGC 3147 and NGC 3998) which have X-ray column densities that are more than an order of magnitude less than we would expect from their optical extinction. Although the spectral fit for NGC 2782 is somewhat uncertain in the soft X-ray band (see Appendix A), the X-ray absorption affecting the other three sources is very well constrained.

Figure 2.12 (left) shows the X-ray spectrum of NGC 3147 with the absorbing column density of $1.217 \times 10^{20}$ cm$^{-2}$ obtained from a free fit. On the right-hand-side of the same figure $N_{\rm H}$ has been fixed to $3.074 \times 10^{21}$ cm$^{-2}$, which is the value we would expect from applying the relationship of Bohlin et al. (1978) to the colour excess ($E(B-V) = 0.53$). The higher value of $N_{\rm H}$ provides a poor fit to the data

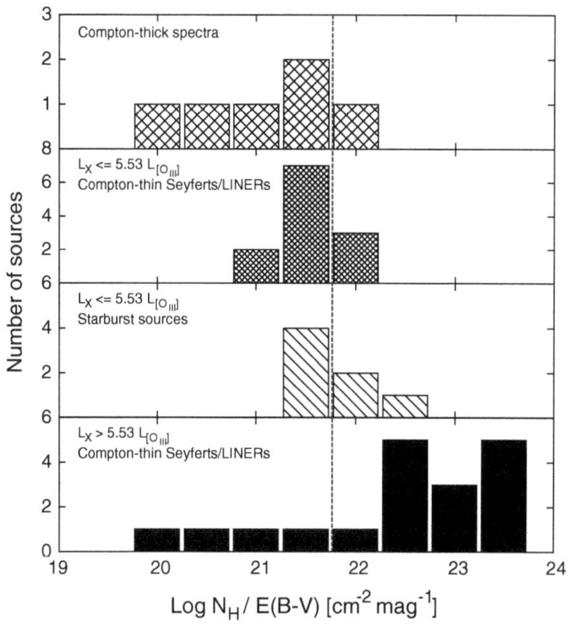


**Fig. 2.11** Histogram showing the ratio of column density ($N_H$) to colour excess ($E(B\text{-}V)$, Ho et al. 1997a). *Top panel* Sources with X-ray spectra that show evidence of Compton thickness. *Second panel* AGN (Seyferts and LINERs) with low X-ray to [OIII] ratios and X-ray spectra that do not show evidence of Compton thickness. The only type-1 Seyfert in this population is NGC 1275, the central member of the Perseus cluster. *Third panel* Starburst sources (H II/transition regions). *Bottom panel* AGN with high X-ray to [O III] ratios, most of which are classified as type-1 Seyferts or LINERs. The *vertical red line* at $5.8 \times 10^{21}$ $cm^{-2}$ $mag^{-1}$ shows the known empirical relationship between column density and colour excess (Bohlin et al. 1978). Seyferts/LINERs with high X-ray to [O III] ratios generally show much higher absorption in the X-ray band than in the optical, presumably due to the origin of the X-rays being deep within the AGN where the photons are subjected to much greater obscuration than that which affects the *narrow line* region further out. Seyferts/LINERs with low X-ray to [O III] ratios instead show $N_H/E(B-V)$ ratios that are consistent with absorption by the same material in both X-ray and optical bands

at low energies and the source therefore exhibits much lower extinction in the X-ray band than in the optical. Clearly this discrepancy between the X-ray and optical observations of this source cannot be attributed to uncertainty in fitting the X-ray absorption component.

The X-ray emission from AGN is generally believed to come from the innermost regions of the system and, even if the X-ray source in NGC 3147 is unobscured within the inner few parsecs, we would still expect there to be absorption from the same material (perhaps the NLR) that is responsible for the reddening of the Balmer lines. Of course, it is possible that the narrow Balmer lines that are used to estimate $E(B-V)$ may not originate from the vicinity of the central black hole, but the *Chandra* ACIS image shown in Appendix B fails to identify any other strong, compact X-ray sources within a radius of 30 arcseconds of the central source (about 6 kpc at

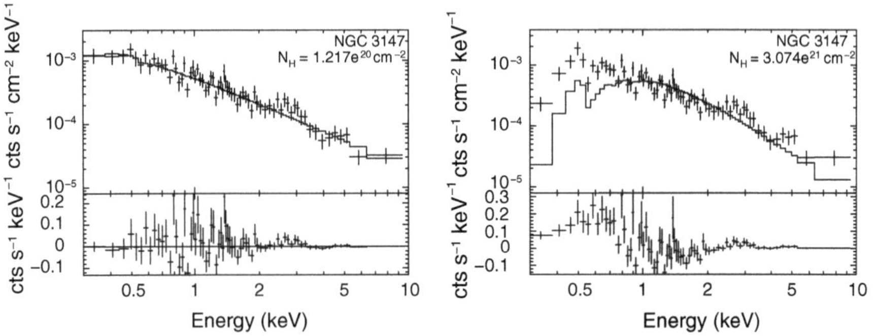


**Fig. 2.12** *Chandra* spectrum of the Seyfert-2 galaxy NGC 3147. *Left* Column density, $N_H$, is allowed to vary during the fit. The best-fit value is $N_H = 1.217 \times 10^{20}$ cm$^{-2}$, which is more than an order of magnitude less than the value we would expect from the optical extinction. *Right* $N_H$ is fixed to $3.074 \times 10^{21}$ cm$^{-2}$, which is the value we would expect from applying the relationship of (Bohlin et al. 1978) to the colour excess ($E(B - V) = 0.53$). This larger column density provides a very poor fit to the data at low energies

an estimated distance of 40.9 Mpc) that could be responsible for the ionisation. A more likely explanation is that the NLR around NGC 3147 is anisotropic, and our line of sight to the central black hole passes through a region of the NLR that is less dense.

NGC 3147 is an example of a class of galaxy often referred to as a 'true' Seyfert-2 (e.g. Ptak et al. 1996; Panessa and Bassani 2002; Georgantopoulos and Zezas 2003; Matt et al. 2012; Bianchi et al. 2008, 2012), which is a term used to describe a Seyfert nucleus that appears unobscured and in which no broad lines are detectable in the optical spectrum. In such cases, it is expected that the BLR must be missing as the absorption is insufficient to entirely obscure these lines. Nicastro et al. (2003) suggested a mechanism whereby the BLR is formed by accretion disc instabilities and showed how it may therefore disappear at accretion rates lower than $\dot{m}_E \leq 10^{-3}$, a limit which is consistent with the low Eddington ratio of NGC 3147 ($\lambda = 2.811 \times 10^{-4}$). Contrary to this claim, however, I find many of the Palomar sources with broad emission lines (especially amongst the LINER population) have Eddington ratios well below the $10^{-3}$ limit (see Fig. 2.6), although it must be added that the detection of broad H$\alpha$ and H$\beta$ lines in the spectra of the low accretion rate sources (Ho et al. 1997c) is likely to be subject to some considerable uncertainty.

Various alternative theories have been proposed to explain the lack of broad lines in NGC 3147. For instance, Matt et al. (2012) argue that the source could actually be Compton thick (despite the soft, power-law spectrum suggesting otherwise) if a highly ionised surface were responsible for the reflection. However, the authors' *Suzaku* observation of energies above 10 keV failed to find any convincing evidence of an obscured AGN. A more likely explanation for the lack of broad lines is possibly that NGC 3147 is completely face-on in terms of inclination (as suggested by Shi et al. (2010)), and the orbital motion which we would normally expect to broaden the emission lines is therefore perpendicular to our line of sight.

## 2.13 X-ray Luminosity, Radio Luminosity and the Fundamental Plane

The emission from AGN is distributed throughout the electromagnetic spectrum and, in order to understand the nature of LLAGN, it will be necessary to consider more than one waveband. The penetrating nature of radio emission, combined with the very high angular resolution that can be obtained, makes it particularly suitable for studying AGN. In XRB systems it is known that radio loudness increases at lower accretion rates as the source enters the hard state, which is indicative of the presence of jets launched by a geometrically thick, radiatively inefficient accretion flow (e.g. Meier 2001; Wu and Cao 2008). Merloni et al. (2003) found a relation between the black hole mass, the radio luminosity and the X-ray luminosity (known as the fundamental plane) such that $\log L_{\rm R} = (0.60^{+0.11}_{-0.11})\log L_{\rm X} + (0.78^{+0.11}_{-0.09})\log M_{\rm BH} + 7.33^{+4.05}_{-4.07}$. This relationship applies for both AGN and stellar mass black holes and could provide a powerful method of confirming the AGN nature of many LLAGN.

Figure 2.13 (top left) shows a plot of X-ray luminosity against radio luminosity, where the X-ray luminosity data are from my *Chandra* survey and the radio data are from the literature (the Palomar sources from Nagar et al. (2005) and the PG quasars from Kellermann et al. (1994)). The plot shows considerable scatter, with the Seyfert-1s clearly distinct from the LINERs. By fitting a fundamental plane to the data, and adjusting the $x$-axis accordingly, the scatter is reduced and the correlation becomes much tighter (see Fig. 2.13, top right). From my fits I find that $\log L_{\rm R} = 0.68\log L_{2-10\,\rm keV} + 0.92\log M_{\rm BH} + 3.34$, which compares favourably to the values found by Merloni et al. (2003). However, the same cannot be said of the fits to only the Palomar sources ($\log L_{\rm R} = 0.37\log L_{2-10\,\rm keV} + 0.77\log M_{\rm BH} + 16.57$) or the LINER sources ($\log L_{\rm R} = 0.48\log L_{2-10\,\rm keV} + 0.77\log M_{\rm BH} + 12.35$— see Fig. 2.13, bottom left and right). Unfortunately, very few of the H II regions or transitions regions have radio luminosities available in the literature (although, in the case of the former, very few are expected to have compact radio cores). If these data were available then the fundamental plane could potentially be used to verify whether or not the emission in these sources is of an AGN origin. For example, if the X-ray and radio emission were caused by a compact population of 1,000 X-ray binary systems in the vicinity of a $10^6\,M_\odot$ SMBH then the radio emission would be expected to be $\sim$2.7 orders of magnitude weaker than if the emission was from the SMBH itself. A source with these properties would be sufficiently offset from the fundamental plane correlation to be easily identifiable. The forthcoming LeMMINGs e-Merlin legacy survey will provide detailed radio observations of all Palomar objects above declination 20° and may help provide some answers concerning the nature of these sources.

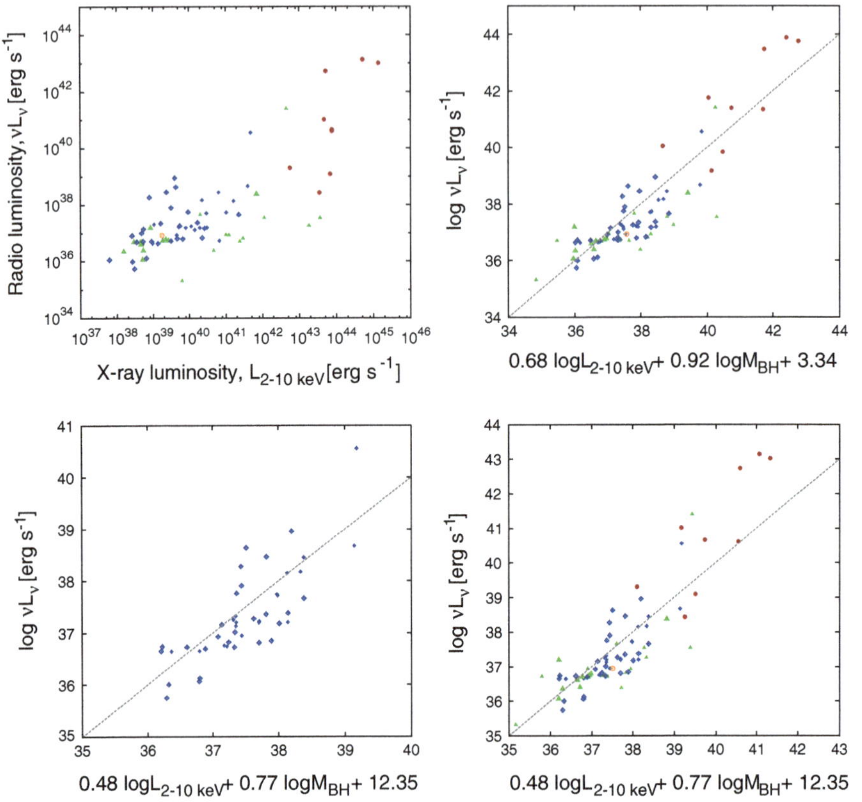


**Fig. 2.13** *Top left* Plot of X-ray luminosity ($L_{2-10\ \mathrm{keV}}$) versus radio luminosity ($\nu L_{\nu}$). *Top right* As before, except the X-ray luminosity has been corrected by fitting a fundamental plane to the data, such that $\log L_{\mathrm{R}} = 0.68 \log L_{2-10\ \mathrm{keV}} + 0.92 \log M_{\mathrm{BH}} + 3.34$. *Bottom left* Best fit to the fundamental plane from my LINER data. I find that $\log L_{\mathrm{R}} = 0.48 \log L_{2-10\ \mathrm{keV}} + 0.77 \log M_{\mathrm{BH}} + 12.35$. *Bottom right* The fundamental plane fitted to the LINER data is applied to the other sources as well, and appears to be an unsatisfactory fit to the quasars and the two radio-loud Palomar sources (NGC 315 and NGC 1275). In all plots the *filled* and *open blue diamonds* represent type-1.0–1.9 and type-2 LINERs respectively, filled and *open green diamonds* represent type-1.0–1.9 and type-2 Seyferts respectively, *red circle* represent the PG quasars and the *orange circles* represent the transition regions. X-ray data are from my *Chandra* survey, radio data for the Palomar sources are 15 GHz VLA observations (Nagar et al. 2005) and radio data for the PG quasars are 5 GHz VLA observations (Kellermann et al. 1994)

## 2.14 Summary of the Main Observational Results

- The 2–10 keV X-ray spectra of Palomar galaxies show no indication of Compton thickness amongst sources where the luminosity of the [O III] line satisfies $L_{[\mathrm{O\,III}]} \leq 10^{39}\ \mathrm{erg\,s^{-1}}$. At low luminosities the ionisation of the narrow line region may be dominated by stellar processes and the ratio of optical line strength to X-ray luminosity is probably unreliable as a Compton-thick indicator.

- Seyferts and LINERs with high X-ray to [O III] ratios show (by their $N_H/E(B-V)$ ratios) much higher absorption in the X-ray band than in the optical, indicating that the origin of the X-ray emission is deep within the AGN where the photons are subjected to much greater obscuration that that which affects the narrow line region further out. Seyferts/LINERs with low X-ray to [O III] ratios instead show $N_H/E(B-V)$ ratios which are consistent with absorption by the same material in both the X-ray and optical bands.
- The majority of type-2 Seyferts and LINERs in the sample have X-ray column densities of $\sim 10^{21}$ cm$^{-2}$, whereas type-1 Seyferts and LINERs have column densities that are distributed evenly over a large range ($<10^{24}$ cm$^{-2}$). The type-2 sources are therefore consistent with either a non-AGN origin or with scattering of intrinsic AGN emission from a reprocessor further out from the centre.
- H II region nuclei and LINERs all have Eddington ratios below $\sim$0.1 % of the Eddington limit, and show little or no overlap in black hole mass (with the LINERs occupying the high end of $M_{BH}$-space). Some LINERs with broad optical emission lines have Eddington ratios as low as $10^{-6}$, which is not consistent with theories promoting the disappearance of the broad line region at low accretion rates.

## 2.15 Discussion

Low luminosity AGN remain a somewhat untidy, confusing and complex class of object, with no single model seemingly able to describe all the properties in an entirely consistent way. For instance, there's no doubt that the unified model of Antonucci and Miller (1985) successfully explains the presence of polarised broad lines in many sources by invoking the dusty torus but, equally, it's also clear that there are just as many type-2 sources to which this model does not apply. The major difficulty that is encountered when studying low accretion rate sources is that the luminosity of the AGN no longer greatly exceeds that of the other sources expected to exist within the host galaxy, and contamination is therefore a problem at all wavelengths. Even the sub-arcsecond angular resolution of *Chandra* does not necessarily isolate the emission related to the accretion flow. LLAGN may therefore be fairly diverse in nature and, in order to compare these sources to Galactic X-ray binary systems, it is likely that careful selection of the sample will be more effective than attempting to model all the different spectral classifications of LLAGN.

Possibly the most important progression towards an understanding of LLAGN would be to reliably identify the Compton-thick candidates. In this chapter I have shown that the use of the [O III]$_{5007\lambda}$ emission line as an indicator of intrinsic X-ray luminosity is unreliable at low luminosities and, in fact, there is little evidence for Compton thickness amongst any source where $L_{[O\,III]} \leq 10^{39}$ erg s$^{-1}$. Instead, at these low luminosities the ionisation of the narrow line region may be dominated by stellar processes, and the discovery of a compact and massive stellar cluster co-existing with the AGN in the inner few parsecs of the transition region galaxy NGC 4303 (Colina and Gonzalez Delgado 2002; Jiménez-Bailón et al. 2003) raises the possibility that nuclear star formation occurring on scales comparable to the torus may be common.

One thing that is clear, however, is that the detection of broad optical lines implies the presence of an AGN and broad-line sources generally show a tight correlation when their X-ray luminosity is plotted against properties such [O III] luminosity, radio luminosity or H $\alpha$ luminosity. It seems reasonable to expect that studies which are restricted to only type-1 Seyferts and LINERs are likely to find more similarities in behaviour with XRBs than studies which include all spectral types, and the presence of broad lines in some LINER sources at Eddington ratios as low as $10^{-6}$ means that such a restriction would have little effect on the range of accretion rates surveyed.

## References


Antonucci, R. R. J., & Miller, J. S. (1985). Spectropolarimetry and the nature of NGC 1068. *Astrophysical Journal, 297*, 621–632.

Baldwin, J. A., Phillips, M. M., & Terlevich, R. (1981). Classification parameters for the emission-line spectra of extragalactic objects. *Publications of the Astronomical Society of the Pacific, 93*, 5–19.

Bassani, L., Dadina, M., Maiolino, R., Salvati, M., Risaliti, G., della Ceca, R., et al. (1999). A three-dimensional diagnostic diagram for Seyfert 2 galaxies: probing X-ray absorption and Compton thickness. *Astrophysical Journal Supplement, 121*, 473–482.

Baumgartner, W. H., Tueller, J., Markwardt, C. B., Skinner, G. K., Barthelmy, S., Mushotzky, R. F., et al. (2012). The 70 Month swift-BAT all-sky hard X-ray survey. ArXiv e-prints.

Bianchi, S., Corral, A., Panessa, F., Barcons, X., Matt, G., Bassani, L., et al. (2008). NGC 3147: a 'true' type 2 Seyfert galaxy without the broad-line region. In L. Lanteri, C. M. Raiteri, A. Capetti, and P. Rossi, (Eds.), 8th National Conference on AGN.

Bianchi, S., Panessa, F., Barcons, X., Carrera, F. J., La Franca, F., Matt, G., et al. (2012). Simultaneous X-ray and optical observations of true type 2 Seyfert galaxies. *Monthly Notices of the Royal Astronomical Society, 426*, 3225–3240.

Blackburn, J. K. (1995). FTOOLS: a FITS data processing and analysis software package. In R. A. Shaw, H. E. Payne, and J. J. E. Hayes, (Eds.), Astronomical Data Analysis Software and Systems IV, volume 77 of ASP Conference Series, p. 367.

Bohlin, R. C., Savage, B. D., & Drake, J. F. (1978). A survey of interstellar H I from L-alpha absorption measurements II. *Astrophysical Journal, 224*, 132–142.

Brightman, M., & Nandra, K. (2008). On the nature of unabsorbed Seyfert 2 galaxies. *Monthly Notices of the Royal Astronomical Society, 390*, 1241–1249.

Cappi, M. (2006). Relativistic blue- and red-shifted absorption lines in AGNs. *Astronomische Nachrichten, 327*, 1012.

Cardelli, J. A., Clayton, G. C., & Mathis, J. S. (1989). The relationship between infrared, optical, and ultraviolet extinction. *Astrophysical Journal, 345*, 245–256.

Cid Fernandes, R., Schlickmann, M., Stasinska, G., Asari, N. V., Gomes, J. M., Schoenell, W., Mateus, A., & Sodré, Jr., L. (2009). The Starburst-AGN disconnection: LINERs as retired galaxies. In W. Wang, Z. Yang, Z. Luo, and Z. Chen, (Eds.), The Starburst-AGN Connection, volume 408 of Astronomical Society of the Pacific Conference Series, p. 122.

Colina, L., Gonzalez Delgado, R., Mas-Hesse, J. M., & Leitherer, C. (2002). Detection of a super-star cluster as the ionizing source in the low-luminosity active galactic nucleus NGC 4303. *Astrophysical Journal, 579*, 545–553.

Constantin, A., Green, P., Aldcroft, T., Kim, D.-W., Haggard, D., & Barkhouse, W. (2009). Probing the balance of AGN and star-forming activity in the local universe with ChaMP. *Astrophysical Journal, 705*, 1336–1355.

Della Ceca, R., Severgnini, P., Caccianiga, A., Comastri, A., Gilli, R., Fiore, F., et al. (2008). Heavily obscured AGN with BeppoSAX, INTEGRAL, SWIFT, XMM and Chandra: prospects for Simbol-X. *Memorie della Societa Astronomica Italiana, 79*, 65.

Georgantopoulos, I., & Akylas, A. (2010). Comparison between the luminosity functions of X-ray and [OIII] selected AGN. *Astronomy and Astrophysics, 509*, A38.

Georgantopoulos, I., Rovilos, E., Akylas, A., Comastri, A., Ranalli, P., Vignali, C., et al. (2011). On the $L_x$–$L_{6m}$ ratio as a diagnostic for Compton-thick AGN. *Astronomy and Astrophysics, 534*, A23.

Georgantopoulos, I., & Zezas, A. (2003). Chandra observations of NGC 4698: A Seyfert 2 galaxy with no absorption. *Astrophysical Journal, 594*, 704–708.

González-Martín, O., Masegosa, J., Márquez, I., & Guainazzi, M. (2009a). Fitting liner nuclei within the active galactic nucleus family: a matter of obscuration? *Astrophysical Journal, 704*, 1570–1585.

González-Martín, O., Masegosa, J., Márquez, I., Guainazzi, M., & Jiménez-Bailón, E. (2009b). An X-ray view of 82 LINERs with Chandra and XMM-Newton data. *Astronomy and Astrophysics, 506*, 1107–1121.

González-Martín, O., Papadakis, I., Braito, V., Masegosa, J., Márquez, I., Mateos, S., et al. (2011). Suzaku observation of the LINER NGC 4102. *Astronomy and Astrophysics, 527*, A142.

Ho, L. C., Feigelson, E. D., Townsley, L. K., Sambruna, R. M., Garmire, G. P., Brandt, W. N., et al. (2001). Detection of nuclear X-ray sources in nearby galaxies with chandra. *Astrophysical Journal Letters, 549*, L51–L54.

Ho, L. C., Filippenko, A. V., & Sargent, W. L. W. (1997a). A Search for "Dwarf" Seyfert nuclei. III. spectroscopic parameters and properties of the host galaxies. *Astrophysical Journal Supplement, 112*, 315.

Ho, L. C., Filippenko, A. V., & Sargent, W. L. W. (1997b). A search for "Dwarf" Seyfert nuclei. V. demographics of Nuclear activity in nearby galaxies. *Astrophysical Journal, 487*, 568.

Ho, L. C., Filippenko, A. V., Sargent, W. L. W., & Peng, C. Y. (1997c). A Search for "Dwarf" Seyfert Nuclei. IV. Nuclei with Broad H alpha Emission. *Astrophysical Journal Supplement, 112*, 391.

Ho, L. C., Greene, J. E., Filippenko, A. V., & Sargent, W. L. W. (2009). A search for "Dwarf" Seyfert nuclei. VII. A catalog of central stellar velocity dispersions of nearby galaxies. *Astrophysical Journal Supplement, 183*, 1–16.

Hopkins, P. F., Hickox, R., Quataert, E., & Hernquist, L. (2009). Are most low-luminosity active galactic nuclei really obscured? *Monthly Notices of the Royal Astronomical Society, 398*, 333–349.

Inoue, H., Terashima, Y., & Ho, L. C. (2007). Fe K line profile in low-redshift quasars: average shape and eddington ratio dependence. *Astrophysical Journal, 662*, 860–871.

Iyomoto, N., Fukazawa, Y., Nakai, N., & Ishihara, Y. (2001). BeppoSAX observation of NGC 3079. *Astrophysical Journal Letters, 561*, L69–L72.

Jiménez-Bailón, E., Santos-Lleó, M., Mas-Hesse, J. M., Guainazzi, M., Colina, L., Cerviño, M., et al. (2003). Nuclear activity and massive star formation in the low-luminosity active galactic nucleus NGC 4303: Chandra X-ray observations. *Astrophysical Journal, 593*, 127–141.

Kellermann, K. I., Sramek, R. A., Schmidt, M., Green, R. F., & Shaffer, D. B. (1994). The radio structure of radio loud and radio quiet quasars in the palomar bright quasar survey. *Astrophysical Journal, 108*, 1163–1177.

Kewley, L. J., Groves, B., Kauffmann, G., & Heckman, T. (2006). The host galaxies and classification of active galactic nuclei. *Monthly Notices of the Royal Astronomical Society, 372*, 961–976.

Maiolino, R., Salvati, M., Bassani, L., Dadina, M., della Ceca, R., Matt, G., et al. (1998). Heavy obscuration in X-ray weak AGNs. *Astronomy and Astrophysics, 338*, 781–794.

Marconi, A., Risaliti, G., Gilli, R., Hunt, L. K., Maiolino, R., & Salvati, M. (2004). Local supermassive black holes, relics of active galactic nuclei and the X-ray background. *Monthly Notices of the Royal Astronomical Society*, *351*, 169–185.

Matt, G., Bianchi, S., Guainazzi, M., Barcons, X., & Panessa, F. (2012). The Suzaku X-ray spectrum of NGC 3147. Further insights on the best "true" Seyfert 2 galaxy candidate. 2012. *Astronomy and Astrophysics*, *540*, A111.

Meier, D. L. (2001). The association of jet production with geometrically thick accretion flows and black hole rotation. *Astrophysical Journal Letters*, *548*, L9–L12.

Merloni, A., Heinz, S., & di Matteo, T. (2003). A fundamental plane of black hole activity. *Monthly Notices of the Royal Astronomical Society*, *345*, 1057–1076.

Nagar, N. M., Falcke, H., & Wilson, A. S. (2005). Radio sources in low-luminosity active galactic nuclei. IV. Radio luminosity function, importance of jet power, and radio properties of the complete Palomar sample. *Astronomy and Astrophysics*, *435*, 521–543.

Narayan, R., Mahadevan, R., & Quataert, E. (1998). Advection-dominated accretion around black holess. In M. A. Abramowicz, G. Bjornsson, and J. E. Pringle, (Eds.), Theory of Black Hole Accretion Disks, p. 148.

Narayan, R., & McClintock, J. E. (2008). Advection-dominated accretion and the black hole event horizon. *New Astronomy Review*, *51*, 733–751.

Nicastro, F., Martocchia, A., & Matt, G. (2003). The lack of broad-Line regions in low accretion rate active galactic nuclei as evidence of their origin in the accretion disk. *Astrophysical Journal Letters*, *589*, L13–L16.

Osterbrock, D. E. (1977). Spectrophotometry of Seyfert 1 galaxies. *Astrophysical Journal*, *215*, 733–745.

Panessa, F., & Bassani, L. (2002). Unabsorbed Seyfert 2 galaxies. *Astronomy and Astrophysics*, *394*, 435–442.

Panessa, F., Bassani, L., Cappi, M., Dadina, M., Barcons, X., Carrera, F. J., et al. (2006). On the X-ray, optical emission line and black hole mass properties of local Seyfert galaxies. *Astronomy and Astrophysics*, *455*, 173–185.

Pappa, A., Georgantopoulos, I., Stewart, G. C., & Zezas, A. L. (2001). The X-ray spectra of optically selected Seyfert 2 galaxies: are there any Seyfert 2 galaxies with no absorption? *Monthly Notices of the Royal Astronomical Society*, *326*, 995–1006.

Predehl, P., & Schmitt, J. H. M. M. (1995). X-raying the interstellar medium: ROSAT observations of dust scattering halos. *Astronomy and Astrophysics*, *293*, 889–905.

Ptak, A., Yaqoob, T., Serlemitsos, P. J., Kunieda, H., & Terashima, Y. (1996). The nature of the low-luminosity active galactic nuclei in NGC 3147. *Astrophysical Journal*, *459*, 542.

Quataert, E. (2001). Low radiative-efficiency accretion flows. In B. M. Peterson, R. W. Pogge, and R. S. Polidan, (Eds.), Probing the Physics of Active Galactic Nuclei, volume 224 of Astronomical Society of the Pacific Conference Series, p. 71.

Richings, A. J., Uttley, P., & Körding, E. (2011). The connection between radio loudness and central surface brightness profiles in optically selected low-luminosity active galaxies. *Monthly Notices of the Royal Astronomical Society*, *415*, 2158–2172.

Shi, Y., Rieke, G. H., Smith, P., Rigby, J., Hines, D., Donley, J., et al. (2010). Unobscured type 2 active galactic nuclei. *Astrophysical Journal*, *714*, 115–129.

Taniguchi, Y., Shioya, Y., & Murayama, T. (2000). Post-starburst Models of LINERS. *Astronomical Journal*, *120*, 1265–1272.

Terashima, Y., Ho, L. C., & Ptak, A. F. (2000). Hard X-ray emission and the ionizing source in LINERS. *Astrophysical Journal*, *539*, 161–165.

Tran, H. D. (2001). Hidden broad-line Seyfert 2 galaxies in the CFA and 12 $\mu$M Samples. *Astrophysical Journal Letters*, *554*, L19–L23.

Tremaine, S., Gebhardt, K., Bender, R., Bower, G., Dressler, A., Faber, S. M., et al. (2002). The Slope of the black hole mass versus velocity dispersion correlation. *Astrophysical Journal*, *574*, 740–753.

Wu, Q., & Cao, X. (2008). Jet power extracted from ADAF and the applications to X-ray binaries and the radio galaxy FR dichotomy. *Astrophysical Journal*, *687*, 156–161.

Younes, G., Porquet, D., Sabra, B., & Reeves, J. N. (2011). Study of LINER sources with broad H$\alpha$ emission. X-ray properties and comparison to luminous AGN and X-ray binaries. *Astronomy and Astrophysics*, *530*, A149.

Zhang, J. S., Henkel, C., Kadler, M., Greenhill, L. J., Nagar, N., Wilson, A. S., et al. (2006). Extragalactic H_2O masers and X-ray absorbing column densities. *Astronomy and Astrophysics*, *450*, 933–944.

# Chapter 3
# Very Fast X-ray Spectral Variability in Cygnus X-1

*The most exciting phrase to hear in science, the one that heralds new discoveries, is not 'Eureka!' (I found it!) but 'That's funny ...'.*

Isaac Asimov (1920–1992)

## 3.1 Synopsis

In this chapter I switch my focus from AGN to XRBs, and introduce a new technique for examining the rapid variability of the X-ray power-law spectral component. I describe here how I extract and fit spectra from each 0.1 s section of *RXTE* data in order to cross-correlate the count rate with $\Gamma$ at high time resolution, and show how the shape of the cross-correlation function evolves with the hardness ratio of the source. This chapter discusses in detail, and attempts to justify, the methods that are later (see Chap. 4) applied to three XRBs and several AGN.

## 3.2 Introduction

The X-ray spectra of accreting black holes can often be fitted with just a few common components, such as a power-law continuum, an iron K$\alpha$ fluorescence line at $\sim$6.4 keV, a reflection component and, if required, a thermal component possibly representing emission from an accretion disc. The power-law component, which takes the form $N_\gamma \propto E^{-\Gamma}$ (where $\Gamma$ is the photon index, $N_\gamma$ is the photon number density and $E$ is the energy), is usually attributed to Comptonisation of lower energy seed photons by high energy electrons (Sunyaev and Titarchuk 1980) in either a central

---

*This chapter has been published in MNRAS:*
Skipper et al. (2013), MNRAS 434 574-584



C. Skipper, *Fast Spectral Variability in the X-ray Emission of Accreting Black Holes*, Springer Theses, DOI 10.1007/978-3-319-09587-5_3

corona or the base of a jet and is observed in both the soft and the hard spectral states.

The correlation between the power-law photon index ($\Gamma$) and the source accretion rate ($\dot{m}_E$) has already been discussed in the introduction (Sect. 1.4.5) and will be briefly recapped here. Previous studies (e.g. Wu and Gu 2008; Gu and Cao 2009; Sobolewska et al. 2011; Zdziarski et al. 2002) have established a positive correlation between $\dot{m}_E$ and $\Gamma$ (the source is softer when brighter) at accretion rates above a critical accretion rate ($\dot{m}_{crit}$—typically 0.5–1 % of the Eddington limit, $\dot{m}_{Edd}$), and an anti-correlation at accretion rates less than $\dot{m}_{crit}$ (the source is harder when brighter).

In all Comptonisation models the way in which $\Gamma$ varies with luminosity, or accretion rate, is a major diagnostic of the physical parameters of the emission region, such as the temperature and optical depth of the high energy electrons and the energy and flux of the seed photons. These parameters, in turn, depend upon the origin of the seed photons and Comptonising electrons, the latter of which may be of uniform energy, or may be energy stratified as a function of radius.

In the soft state the accretion flow is believed to take the form of a radiatively efficient, optically thick, geometrically thin disc at all radii, with the seed photon supply being dominated by black body photons from the inner edge of the accretion disc. The origin of the positive correlation between $\Gamma$ and $\dot{m}_E$ in the soft state is therefore often attributed to Compton cooling of the corona by seed photons from the accretion disc. In the hard state the flow at small radii may become a geometrically thick but optically thin and radiatively inefficient accretion flow (RIAF) (Gu and Cao 2009; Wu and Gu 2008; Done and Gierliński 2006; Narayan and McClintock 2008; Qiao and Liu 2010; Niedźwiecki et al. 2012) which contributes significantly to seed photon supply through synchrotron and bremsstrahlung emission. The anti-correlation found in the hard state at low accretion rates, when the spectrum is expected to be dominated by Comptonisation of these synchrotron and bremsstrahlung photons, may be due to an injection of high energy electrons into the corona producing both an increase in seed photon emission and a harder Compton-scattered spectrum (Qiao and Liu 2013). However, under the truncated disc model (Esin et al. 1997), the inner radius of the accretion disc in the hard state moves inwards with increasing luminosity, and the balance of seed photon supply may consequently shift from synchrotron and bremsstrahlung emission from the RIAF at low $\dot{m}_E$ to disc black body emission at high $\dot{m}_E$.

Within individual XRBs, however, there is little information available on the relationship between $\Gamma$ and $\dot{m}_E$ on very short time-scales (i.e. comparable, when scaled by mass, with the relatively short timescales on which X-ray spectral variability has been studied in AGN). Wu et al. (2010) showed that the positive correlation exists on time-scales of 62 ms in a single observation of Cygnus X-1, but had to correct for a significant amount of Poisson noise in order to do so. This time-scale corresponds approximately to the Compton cooling time-scale for the corona around a $10\,M_\odot$ black hole (Ishibashi and Courvoisier 2012) or the viscous time-scale and so is one of the shortest time-scales on which observable changes in the emission region may be expected to occur. Scaled by mass, this time-scale also corresponds to approximately 17 h for an AGN with a $10^6\,M_\odot$ black hole and thus enables the study of spectral variability in binaries on almost the shortest (scaled) time-scale on which spectral variability in AGN can be studied. However, XRB systems offer the additional benefit of allowing me to study a much wider range of accretion rates within a single object, including transition across the $\dot{m}_{\rm crit}$ boundary, something which is impossible with an AGN.

In order, therefore, to measure the important $\Gamma$–$\dot{m}_{\rm E}$ diagnostic in a single black hole system, as it passes from the hard to the soft spectral state, across the $\dot{m}_{\rm crit}$ boundary, I select here fifteen observations from the *Rossi X-ray Timing Explorer* (*RXTE*) archive of the well known high mass X-ray binary (HMXB) system Cygnus X-1. This source was chosen because it has comprehensive archival coverage and is sufficiently bright to allow time-resolved X-ray spectroscopy on time-scales as short as 16 ms. In order to study small changes in $\Gamma$ on these very short time-scales I measure $\Gamma$ from 2 million separate observational sections (of time-scales between 16 and 100 ms, depending on flux level) and thus produce a $\Gamma$ lightcurve which I cross-correlate with the count rate lightcurve (used as an indicator of the accretion rate). Thus I can determine not only whether $\Gamma$ rises or falls with count rate but also examine whether or not changes in one of these parameters lags those in the other. In addition, I can measure the characteristic time-scale of the correlation (the width of the cross-correlation function) and examine how this evolves with accretion rate and hardness ratio over much longer time-scales.

## 3.3 Observations

All data used in this work have been selected and retrieved from the *RXTE* archive, which provides a vast repository of observations of Cygnus X-1. To simplify the reduction I have decided to use only those datasets that provide both Proportional Counter Array (PCA) standard-2 and generic binned array mode with at least 16 ms time resolution. The standard-2 data provide good spectral resolution, with 47 channels available to cover the 3–20 keV energy range, but poor time resolution with data binned into intervals of 16 s. In contrast, the generic binned array modes provide excellent time resolution of typically 4–16 ms and the reduced spectral resolution helps to ensure a healthy number of counts per bin.

In total, 15 observations of Cygnus X-1 were retrieved from the *RXTE* archive ranging in date from March 1996 until February 1998. These datasets, listed in Table 3.1, were selected to provide coverage of both the soft-state outburst in 1996 and the more steady hard state in which the source was found for the subsequent two years.

**Table 3.1** A summary of the observations

| No. | Observation | Date | Spectral state | No.[a] Spectra |
|---|---|---|---|---|
| 1 | 10238-01-08-00 | 26 Mar 1996 | Hard | 58,540 |
| 2 | 10238-01-04-00 | 07 Apr 1996 | Intermediate | 30,000 |
| 3 | 10412-01-02-00 | 22 May 1996 | Soft | 25,000 |
| 4 | 10412-01-01-00 | 23 May 1996 | Soft | 32,100 |
| 5 | 10412-01-05-00 | 11 Aug 1996 | Soft | 25,000 |
| 6 | 10241-01-01-00 | 24 Oct 1996 | Hard | 34,972 |
| 7 | 10236-01-01-02 | 17 Dec 1996 | Hard | 102,500 |
| 8 | 20175-01-02-00 | 04 Oct 1997 | Hard | 36,980 |
| 9 | 30157-01-01-00 | 11 Dec 1997 | Hard | 19,100 |
| 10 | 30157-01-03-00 | 24 Dec 1997 | Hard | 28,800 |
| 11 | 30157-01-04-00 | 30 Dec 1997 | Hard | 15,400 |
| 12 | 30157-01-05-00 | 08 Jan 1998 | Hard | 20,000 |
| 13 | 30157-01-07-00 | 23 Jan 1998 | Hard | 25,600 |
| 14 | 30157-01-08-00 | 30 Jan 1998 | Hard | 11,620 |
| 15 | 30157-01-10-00 | 13 Feb 1998 | Hard | 28,692 |

Spectra were classified as either hard, soft or intermediate state depending upon the photon index and whether or not a disc black body component was required to fit the data in the 3–20 keV range

[a] The number of 100 ms spectra extracted from the observation

## 3.4 Spectral Fitting

### *3.4.1 Time-Averaged Fits to Each Complete Observation*

In order to examine variability on time-scales of 100 ms or less, I use the PCA generic binned array mode data. However, there are not enough counts, or spectral resolution, in each 16–100 ms section of these data to allow me to fit a complex model with all parameters free. I must therefore determine, from the lower time resolution, but higher spectral resolution, standard-2 data, a realistic model which can be applied to all the shorter time-scale spectral fits.

Initially, spectra were extracted from the standard-2 datasets for each full observation and fitted in XSPEC version 12.8 over the 3–20 keV energy range with a model consisting of absorption (PHABS), a power-law component and its associated reflection (PEXRAV; Magdziarz and Zdziarski 1995), a Gaussian representing the iron K$\alpha$ fluorescence line at $\sim$6.4 keV (GAUSSIAN) and, if required, a disc black body component (DISKBB). The inclination angle of the PEXRAV component is fixed to 30°, and the fold energy, redshift, metal abundance and iron abundance to default values. My choice of inclination angle is reasonably consistent with both the most recently published figure of $27.1 \pm 0.8°$ (Orosz et al. 2011) and also my own fits to the data, which vary by observation but tend to give best fit values slightly above 30° in the hard state. However, when the inclination angle was fixed to a higher value of 53° I found no discernible differences in the appearance of my count rate versus $\Gamma$ cross-correlation functions, and therefore suggest that the short time-scale, relative changes to $\Gamma$ such as I discuss in this chapter are insensitive (or only weakly sensitive) to the choice of inclination angle.

In four of my fifteen observations (2, 3, 4 and 5) it was found that the addition of a disc black body component with an average disc temperature of 0.47 keV significantly improved the fit. Hereafter, I refer to these four observations as being in the soft or intermediate states and the remaining eleven observations, that lacked an obvious thermal component, as being in the hard state. The soft/intermediate-state observations were also found to have broader average iron K$\alpha$ lines than the hard states (one sigma Gaussian widths of 1.31 and 0.52 keV respectively) and stronger reflection scaling factors (1.66 and 0.46 respectively).

The results of my spectral fitting are shown in Fig. 3.1, which gives $\Gamma$, the hardness ratio ($L_{5-13\,\mathrm{keV}}/L_{3-5\,\mathrm{keV}}$) and the ratio of the 3–20 keV luminosity to the Eddington luminosity. The validity of the latter as a proxy for accretion rate is reduced somewhat by the potentially large differences in bolometric correction between the spectral states, and no attempt has been made here to estimate these. Furthermore, the sensitivity range of the PCA instrument does not provide coverage of the soft (<2 keV) or very hard (>60 keV) energy bands in which most of the power is emitted in the soft and hard states respectively (Zdziarski et al. 2002;

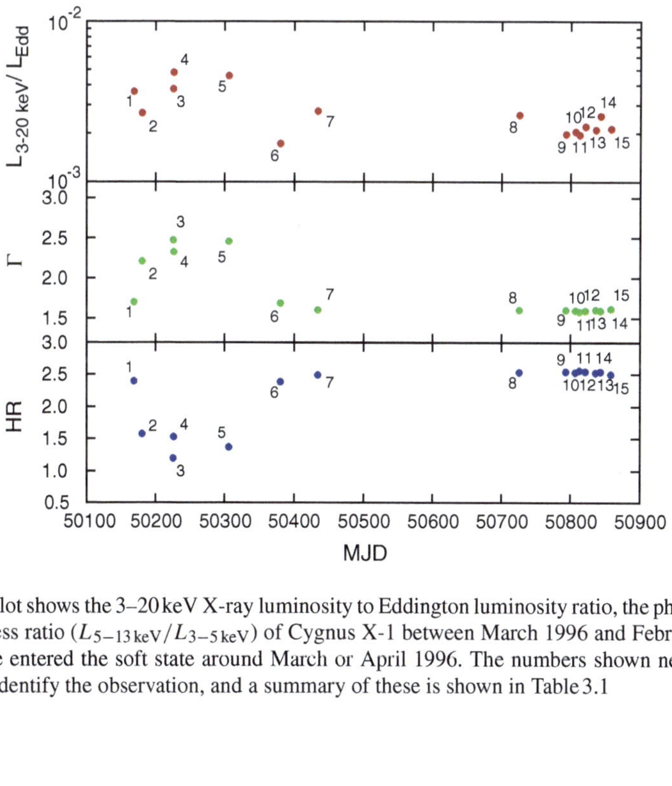


**Fig. 3.1** Plot shows the 3–20 keV X-ray luminosity to Eddington luminosity ratio, the photon index and hardness ratio ($L_{5-13\,\mathrm{keV}}/L_{3-5\,\mathrm{keV}}$) of Cygnus X-1 between March 1996 and February 1998. The source entered the soft state around March or April 1996. The numbers shown next to each datapoint identify the observation, and a summary of these is shown in Table 3.1

Churazov et al. 2001), and the 3–20 keV luminosity may not therefore be an ideal choice for estimating accretion rate. The luminosity and Eddington limits were estimated using a distance of $1.86^{+0.12}_{-0.11}$ kpc (Reid et al. 2011) and a black hole mass of $14.8 \pm 1.0\,M_{\odot}$ (Orosz et al. 2011). The numeric labels correspond to the observation number as shown in Table 3.1. During the 1996 soft-state outburst the increase in $\Gamma$ and the reduction in the hardness ratio is clearly evident.

### *3.4.2 Determining which Model Parameters to Fix*

Although I now have suitable models for each of my observations, there are still not enough counts, or spectral resolution, on 100 ms time-scales to allow all the model parameters to remain free during the fit. Fortunately, I expect that the power-law will be varying more on short time-scales than the other components so therefore adopt a strategy whereby the power-law is allowed to vary and all the other components remain fixed. I must be cautious, however, as although my models show that an iron line and a reflection component are important parts of the overall spectrum, they do not tell me whether the reflection scaling factor or the total flux in the reflected component remains constant over the whole observation. Similarly, I do not know if the flux in the iron line or the equivalent width of the iron line remains constant. Incorrectly fixing either one of these parameters, and particularly the reflection component, can lead to erroneous conclusions regarding the variation of $\Gamma$ with flux.

In order to determine which parameters in my models should be fixed, I extracted 657 spectra—each 16 s in length—from the standard-2 data of the hard-state observation 7, sorted them into order of count rate and then summed them to produce 33 new spectra, each 320 s in length. These data were fitted in XSPEC with *all* components allowed to vary. The results are shown in Fig. 3.2 and reveal that the equivalent width of the iron K$\alpha$ line remains roughly independent of the count rate (which has also been observed by Maccarone and Coppi (2002) for Cygnus X-1 and Nowak et al. (2002) for GX 339-4) and therefore fixing the equivalent width whilst fitting my 100 ms spectra seems entirely reasonable. To do this, I replace the GAUSSIAN component in my models with the GABS component, and fix the line depth to a negative value in order to model an emission line. The reflection scaling factor, however, tends to increase slightly with the count rate (which is likely to be a short time-scale expression of the correlation between reflection and $\Gamma$ reported by Zdziarski et al. 1999), although not to the same extent as the total reflection flux. Therefore, I determine that fixing the reflection scaling factor seems more acceptable than fixing the whole reflection component.

These observations suggest an origin for the iron line, and most of the reflected component, in a close reprocessor such as the accretion disc. It is not clear why there is a small variation in the reflection scaling factor with flux. The positive sense of the correlation is opposite to that predicted by the light bending models designed to explain spectral variability (e.g. Miniutti and Fabian 2004) and, if interpreted within the context of light bending, would imply that the X-ray source becomes closer to

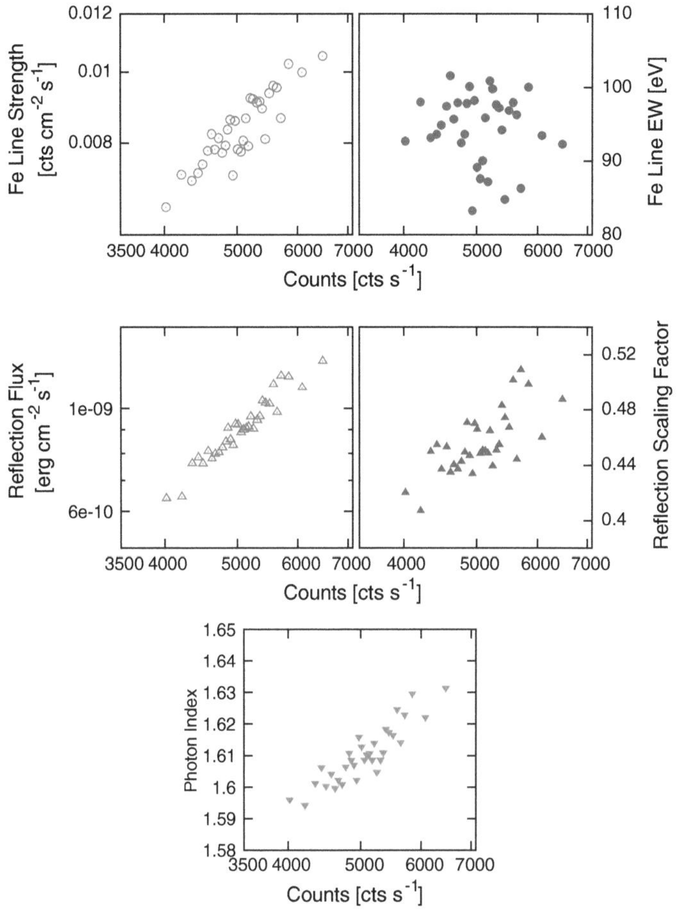


**Fig. 3.2** Relationship between the count rate and the properties of three of the spectral components for the hard-state observation 7 (1996 December 17). A total of 657 spectra, each 16 s in length, were extracted from the standard-2 data and sorted into order of count rate. The 16 s spectra were then summed to produce 33 new spectra each of length 320 s, which offer sufficiently high count rates to allow all the spectral components to vary during the fit. *Top* The strength of the iron K$\alpha$ line at 6.4 keV, expressed in terms of its photon flux (*left*) and equivalent width (*right*). The photon flux scales with the count rate in the whole spectrum whereas the equivalent width remains roughly constant, suggesting that the latter should be fixed when fitting much shorter time-scale spectra. *Centre* The strength of the reflection component expressed in terms of its 3–20 keV flux (*left*) and its reflection scaling factor from the PEXRAV component (*right*). Both properties scale with the count rate (the latter to a lesser extent than the former) which could be indicative of either an increase in the optical depth of the disc or a reduction in the truncation radius at higher accretion rates. Zdziarski et al. (1999) found a similar correlation between reflection and $\Gamma$ in a *Ginga* sample of Seyferts and XRBs. *Bottom* The photon index of the power-law shows a strong correlation with the count rate

the reflector as the overall luminosity goes up. Instead, this correlation may perhaps be attributed to a decrease in the disc truncation radius with increasing flux, which consequently increases the surface area of the reflector. Since I expect the variations in the fitted iron line equivalent width will be dominated by Poisson variations, rather than real ones, I hold the equivalent width fixed in order not to let random variations affect the other parameters.

### *3.4.3 Fitting the Spectra at 100 ms Time Resolution*

Spectra were then extracted at 100 ms time resolution from the binned array mode datasets of the observations listed in Table 3.1, and fitted in XSPEC over the 3–20 keV energy range. The absorption, reflection scaling factor, disc black body component and the equivalent width of the iron K$\alpha$ line were fixed at the values determined from fits to each total observational dataset, and only the photon index and the

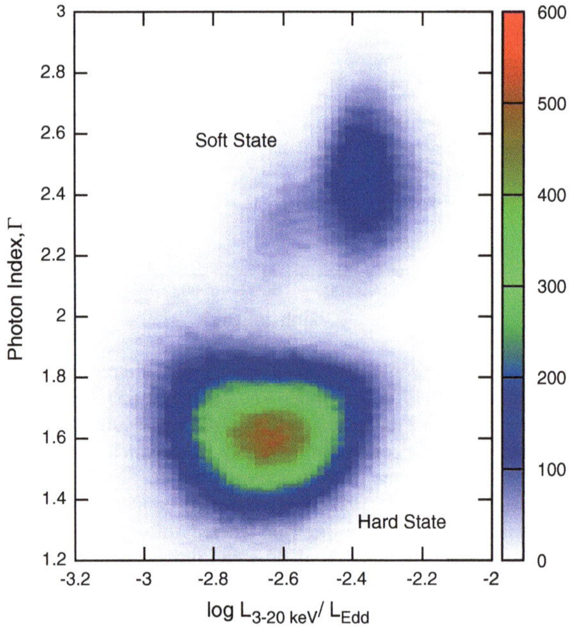


**Fig. 3.3** Density plot of ~500,000 spectra, each 100 ms in length, extracted and fitted from 11 hard-state and 4 soft/intermediate-state observations of Cygnus X-1. The plot shows the ratio of the X-ray luminosity to the Eddington luminosity ($L_{3-20\,\mathrm{keV}}/L_{\mathrm{Edd}}$) against the photon index ($\Gamma$). The hard and soft states are easy to distinguish with the latter forming a trail out towards the *top* of the plot. The scale on the *right-hand-side* refers to the number of spectra per pixel, where each pixel extends over a range $\Delta \log L = 0.02$ on the x-axis and $\Delta\Gamma = 0.01$ on the y-axis. It should be noted that the relative densities of the soft and hard states are biased by my choice of observations, and are not necessarily indicative of the amount of time the source spends in each state

normalisation of the power-law were allowed to vary. The spectra were extracted and fitted automatically by scripts that make use of the HEASARC tools SAEXTRCT and XSPEC. For some of my observations it was also necessary to further rebin the data to improve the number of counts per bin, especially at energies above 12 keV.

The range covered by these spectra is displayed in Fig. 3.3, which shows a density plot of approximately 500,000 spectra, each 100 ms in length, that compares the X-ray luminosity (divided by the Eddington luminosity) with the photon index of the power-law component. The soft and hard states are easy to distinguish, with the former exhibiting a much higher average photon index (2.42, compared to 1.63) and higher luminosity ($L_{3-20}$ keV was 0.392 % of the Eddington limit, compared to 0.224 %). The cluster of points that appear intermediate between the soft and hard states are from observation 2 (7th April 1996), which relates to the start of the soft-state outburst and shows only a weak disc black body component in its X-ray spectrum.

## 3.5 Cross-Correlation of Γ and Count Rate on 100 Millisecond Time-Scales

### *3.5.1 The Hard States*

In Fig. 3.4 I show the cross-correlation functions (CCFs) between Γ and the 3–20 keV count rate for all eleven of my hard-state observations at 100 ms time resolution. I choose to use the discrete correlation function (DCF) method of Edelson and Krolik (1988) and normalise my results using the excess variance. The CCFs in this plot are ordered by hardness ratio ($H.R. = L_{5-13\,\mathrm{keV}}/L_{3-5\,\mathrm{keV}}$), which is also shown on the right hand side of each panel along with the accretion rate in Eddington units. I choose to use the source count rate instead of the flux in my CCFs as the latter, like Γ, is derived from my spectral fits and a degeneracy between these properties could produce erroneous results.

As each measurement of Γ is subject to considerable statistical uncertainty, the peak values of some of the hard-state CCFs are not high. However, a large number of spectra contribute to each CCF (between ~10,000 and ~100,000 for each observation) and the results are none the less highly significant. A further discussion of noise in my data can be found in Sect. 3.8.

The CCFs suggest the presence of positive and negative correlations, both peaking at around zero lag, and either of which can be missing from individual CCFs. Most of the plots reveal a strong lack of symmetry, with the positive correlation declining more slowly towards negative lags (source counts trail the photon index) and the anti-correlation always declining towards positive lags (photon index trails the source counts).

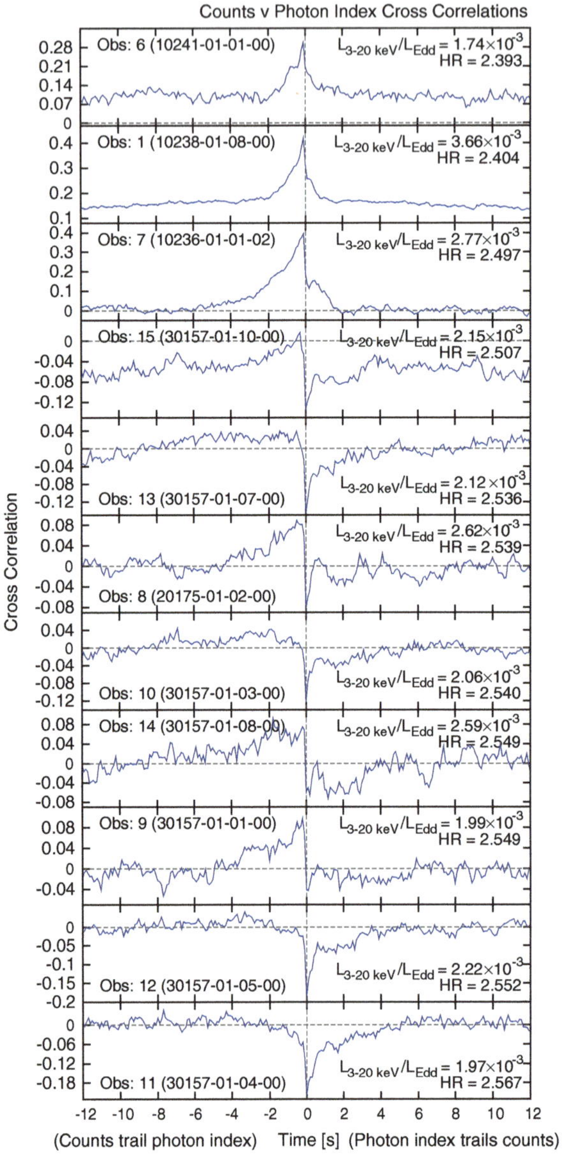


**Fig. 3.4** Cross-correlation functions between the photon index and the 3–20 keV source counts for 11 hard-state observations of Cygnus X-1. The observations are ordered by hardness ratio ($H.R. = L_{5-13\,\text{keV}}/L_{3-5\,\text{keV}}$), which is also shown on the *right hand side* of each panel along with the X-ray luminosity to Eddington luminosity ratio. The positive peaks imply that the source becomes softer when brighter and the negative peaks that the source becomes harder when brighter. The plots show a broad positive correlation and a narrower anti-correlation, the relative strength of which appears to be more strongly connected to the hardness ratio than the accretion rate. Although the strengths of some of the correlations in this plot are weak, they cannot be explained as the result of noise (see Sect. 3.8 for a discussion of this)

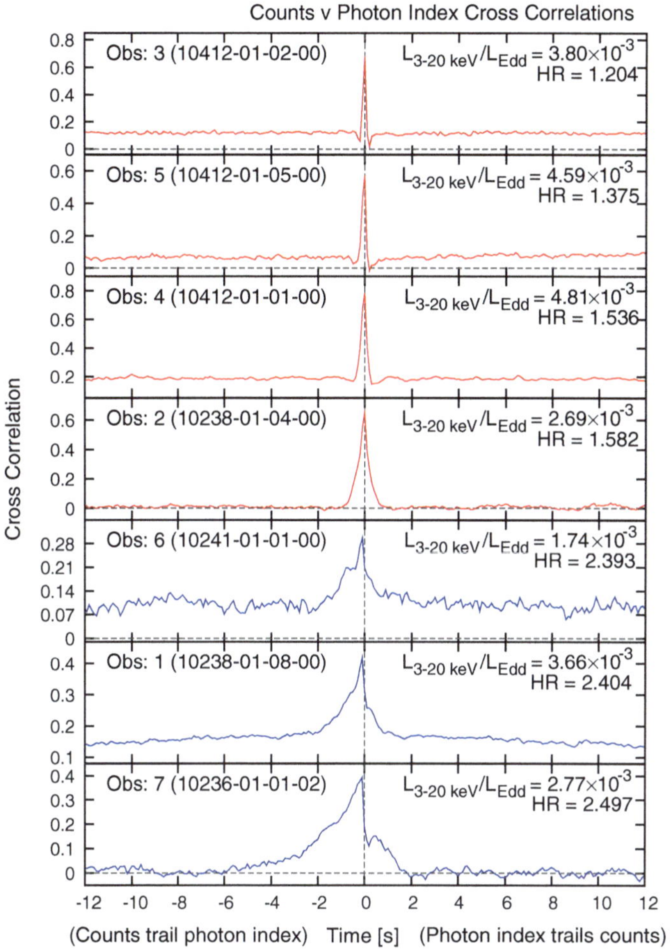


**Fig. 3.5** Cross-correlation functions between the photon index and the 3–20 keV source counts for four soft/intermediate-state observations of Cygnus X-1 (*red lines*) and, shown for comparison, the three hard-state observations with the lowest hardness ratios (*blue lines*). The CCFs are ordered by hardness ratio ($H.R. = L_{5-13\,\mathrm{keV}}/L_{3-5\,\mathrm{keV}}$), which is also shown on the *right hand side* of each panel along with the X-ray luminosity to Eddington luminosity ratio. All the CCFs show a single positive peak, *centred* at or near *zero lag*. Both the width of the peak and its asymmetry appear to increase with hardness ratio and there is no sign of the anti-correlation which is evident in many of the hard-state CCFs

The most obvious conclusion to be drawn from examination of Fig. 3.4 is that there is a steady transition from a positive to a negative correlation (i.e. an anti-correlation) as the hardness ratio increases. Some of the CCFs (particularly from observations 8, 9 and 14) show a greater degree of asymmetry than others, with a positive correlation being found at negative lags and a negative correlation at positive lags. I find that the changes in CCF shape occur over a quite limited range of hardness ratios. Examination of the data in the 2–3 keV range indicates that if the soft band was extended down to 2 keV, the hardness ratio range would expand. However as the response of the PCA below 3 keV is both lower, and more uncertain, than at higher energies, I invoke the relatively standard procedure of only using data down to 3 keV.

The steady transition in the shape of the CCF is not seen so clearly if the observations are ordered by count rate. In particular I find that observation 6, which dates from a time when the source has just emerged from the soft state (24th October 1996), is both the softest and faintest observation and shows no sign of an anti-correlated component in its CCF. These observations suggest that the relative strength of the positive and negative correlations depends more strongly upon the hardness ratio than upon the accretion rate. I note, however, that given the large spectral changes that occur in this source (e.g. Zdziarski et al. 2002) the 3–20 keV count rate is far from being a perfect indicator of accretion rate, being particularly poor during transitions between states.

### *3.5.2 The Soft/Intermediate States*

The cross-correlation functions of photon index against source counts are shown in Fig. 3.5 for my four soft/intermediate-state observations and, shown for comparison, the three hard-state observations with the softest hardness ratios. The soft-state CCFs show less complex profiles than the hard-state CCFs, with much stronger, narrower and more symmetric positive correlations. The anti-correlation that was evident in most of the hard-state CCFs is not obvious here.

**Table 3.2** Properties of the four soft/intermediate-state CCFs, shown along with those of the three hard-state observations with the lowest hardness ratios

| No. | Observation | Date | $L_{3-20\,\mathrm{keV}}/L_{\mathrm{Edd}}$[a] $\times 10^{-3}$ | H.R.[b] | Width (FWHM)[c] (ms) | Width LHS[d] (ms) |
|---|---|---|---|---|---|---|
| 1 | 10238-01-08-00 | 26/03/1996 | 3.67 | 2.404 | 888 | 822 |
| 2 | 10238-01-04-00 | 07/04/1996 | 2.69 | 1.582 | 382 | 240 |
| 3 | 10412-01-02-00 | 22/05/1996 | 3.80 | 1.204 | 140 | 91 |
| 4 | 10412-01-01-00 | 23/05/1996 | 4.81 | 1.536 | 249 | 169 |
| 5 | 10412-01-05-00 | 11/08/1996 | 4.59 | 1.375 | 189 | 138 |
| 6 | 10241-01-01-00 | 24/10/1996 | 1.74 | 2.393 | 895 | 844 |
| 7 | 10236-01-01-02 | 17/12/1996 | 2.82 | 2.497 | 1348[e] | 1352[f] |

In this limited sample I find that the widths of the peaks in the CCFs show a stronger correlation with the hardness ratio than the accretion rate

[a] Ratio of X-ray luminosity to Eddington luminosity

[b] Hardness ratio ($L_{5-13\,\mathrm{keV}}/L_{3-5\,\mathrm{keV}}$)

[c] FWHM of peak

[d] Width of the left-hand side of the peak, measured from the left-hand side of the half maximum up to the y-axis

[e,f] At half maximum the entire peak is to the left of the y-axis, and therefore the distance from the left-hand side to the y-axis is greater than the FWHM for this observation

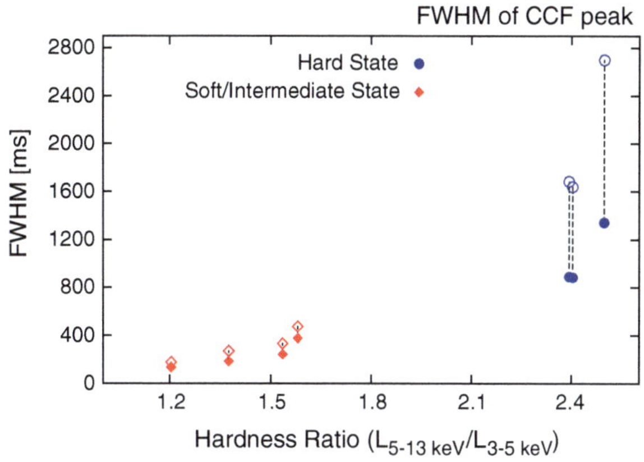


**Fig. 3.6** The widths of the peaks in the CCFs plotted against the hardness ratio for the four soft/intermediate states and the three hard-state observations with the lowest hardness ratios. Each of the peaks was found to be slightly asymmetric, with the width of the *left-hand side* slightly wider than that of the *right*. The symbols in the plot represent both the measured FWHM (*filled symbols*) and twice the width of the *left-hand side* (*open symbols*). In both cases the width is correlated with the hardness ratio, which possibly suggests a connection to the truncation radius of the accretion disc

The widths of the peaks found in the soft/intermediate-state CCFs, along with those found in the hard-state observations 1, 6 and 7, have been measured (see Table 3.2) and the FWHM of the peaks are plotted against the hardness ratios ($L_{5-13\,\mathrm{keV}}/L_{3-5\,\mathrm{keV}}$) in Fig. 3.6. Although this sample is limited in size, my data suggest that the width of the peak is correlated with the hardness ratio of the source but not the accretion rate.

## 3.6 Cross-Correlation of Γ and Count Rate on 16 Millisecond Time-Scales

### *3.6.1 A Closer Look at the Soft State, Bright Hard State and Faint Hard State*

All results discussed up to this point have been based upon 100 ms time-resolution spectra, and separate cross-correlation functions have been generated for each observation. However, it would be useful to examine the differences between the soft states and the bright and faint hard states in a little more detail and, to do this, I approach the problem in two ways.

For the soft states and the bright hard states I choose to reduce the data again, but this time using 16 ms time-resolved spectra. For the faint hard states I do not have sufficient counts to allow me to fit spectra on this time-scale, so I instead choose to combine the existing 100 ms data from several observations and, from this combined data, produce a single cross-correlation function with improved signal-to-noise ratio.

Figure 3.7 (top) shows the CCF for the soft-state observation 4 (1996 May 23) using 16 ms time-resolved spectra. This plot was generated from a total of 199, 989 spectra, with an average of 147 counts in the 3–20 keV range and a mean photon index of 2.30. The peak in the CCF is smooth and featureless even on time-scales as short as these, but I also find a very slight asymmetry that was not clear at 100 ms time resolution.

Figure 3.7 (centre) shows the CCF for the bright hard-state observation 7 (1996 December 17), also using 16 ms time-resolved spectra. This plot was generated from a total of 361,868 spectra, with an average of 66 counts in the 3–20 keV range and a mean photon index of 1.48. The model is the same as that which was used for the soft-state observation, except the disc black body component was not required. Although the peak is found close to zero lag, it is not quite symmetric in appearance and is instead skewed towards negative time lags.

Figure 3.7 (bottom) shows the CCF for the combined faint hard-state observations (8–15), using 100 ms time-resolution spectra. The narrow anti-correlation is clearly evident and is found mostly at positive lags, while the positive correlation is either missing or has been significantly broadened.

The peaks found in the CCFs of Figs. 3.4 and 3.5 show a degree of asymmetry that tends to increase with the hardness ratio. I find that the CCFs that show a positive correlation between source counts and $\Gamma$ often have peaks that are skewed towards negative lags and those showing an anti-correlation are always skewed towards positive lags. A possible explanation for this could be that the positive and negative correlation peaks are actually symmetric in each of my observations, but as the strength of the correlation decreases the presence of another, non-symmetric, component begins to become more apparent, resulting in a peak that is broader on one side of the y-axis than it is on the other.

In Fig. 3.8 (left panel) I plot the autocorrelation functions (ACFs) of the source counts for four soft/intermediate-state and five hard-state observations and find that the widths of the peaks, especially those generated from the soft-state observations, are reasonably consistent with the widths of the peaks in the source counts/photon index CCFs (Fig. 3.8, centre panel). The ACF widths tend to increase with the hardness ratio, which is consistent with reports that the power spectral density (PSD) shifts towards higher frequencies as the spectra soften (see Pottschmidt et al. 2003 and references therein). I then compare the right-hand side of the CCFs with the reflection of the left-hand side (shown as a dotted purple line in Fig. 3.8, centre panel) and, in the rightmost panel of the same figure, plot the difference between the two. The result is not unlike the anti-correlation peak found in some of the hard-state CCFs (Fig. 3.4) and raises the possibility that this highly asymmetric component may be present in *all* of my CCFs, including those from the soft-state observations.

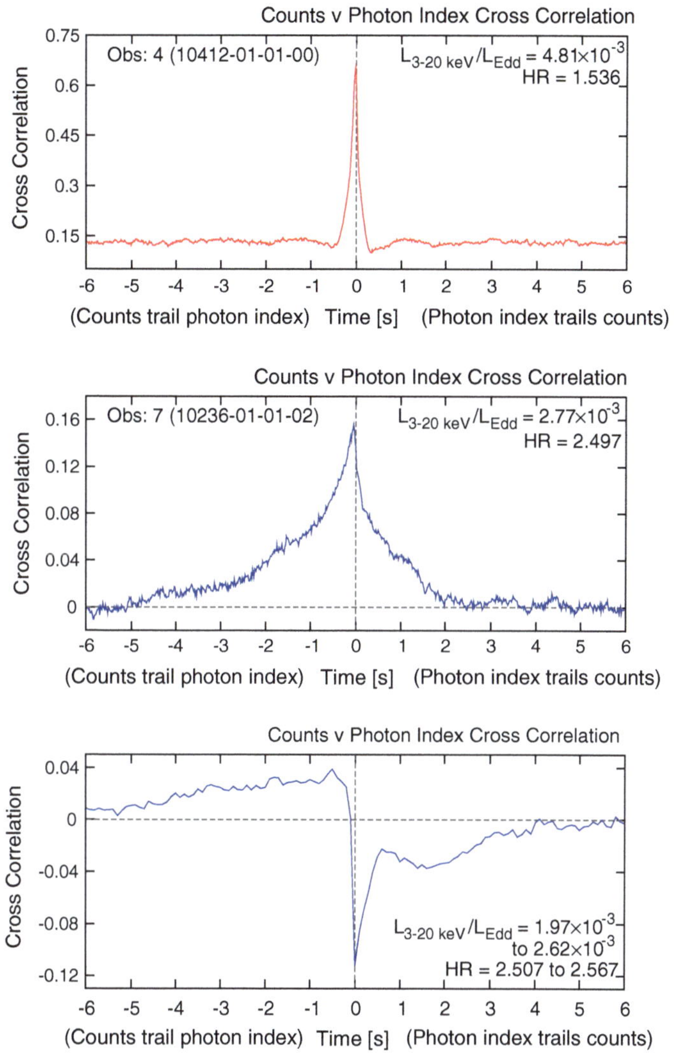


**Fig. 3.7** *Top* Cross-correlation function (CCF) between the photon index and the 3–20 keV source counts for the soft-state observation 4 (1996 May 23), shown with 16 ms time resolution. The peak in the CCF is smooth and featureless, even on time-scales as short as these, but is slightly asymmetric. *Centre* CCF between the photon index and the 3–20 keV source counts for the bright hard-state observation 7 (1996 December 17), shown with 16 ms time resolution. The peak in the CCF is non-symmetric with a gradual decline towards negative time-lags and a steeper decline towards positive time-lags. *Bottom* CCF between the photon index and the 3–20 keV source counts, produced by combining the data from 8 low accretion rate/high hardness ratio observations of Cygnus X-1. The plot reveals a narrow, non-symmetric anti-correlation in which the photon index trails the counts by a period of less than one second. In addition to this peak, I find the two properties are weakly correlated at negative time-lags and weakly anti-correlated at positive time lags

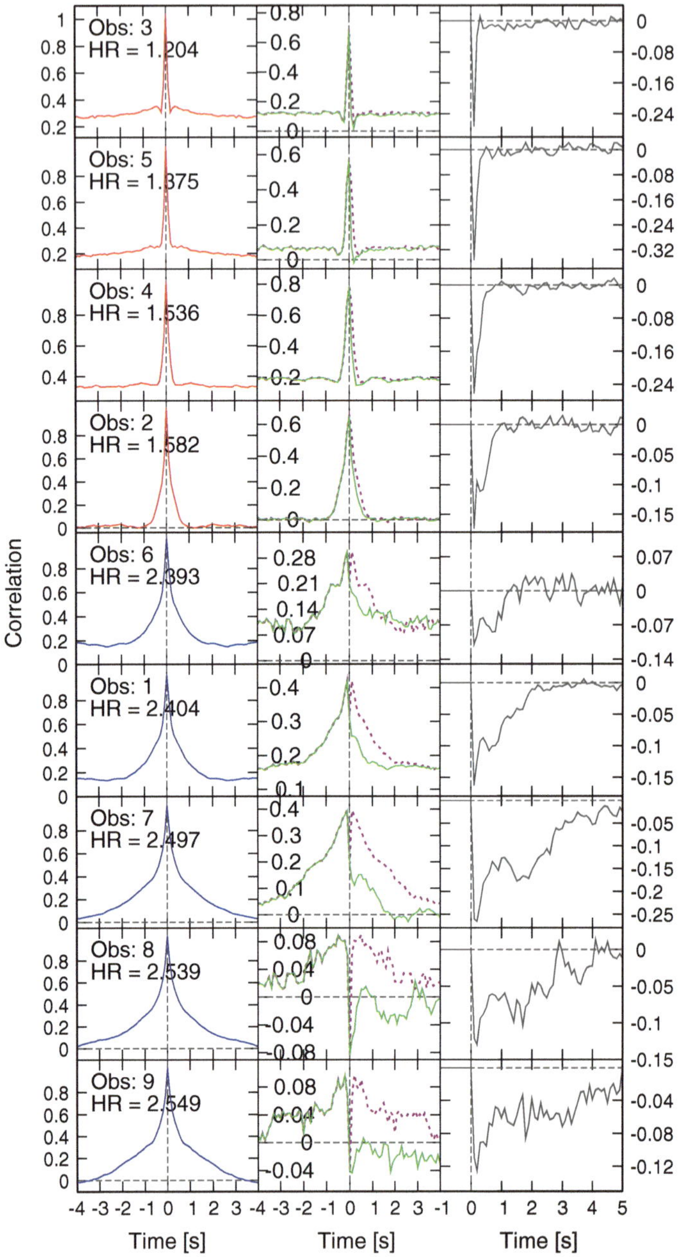
Obs: 3
HR = 1.204
Obs: 5
HR = 1.375
Obs: 4
HR = 1.536
Obs: 2
HR = 1.582
Obs: 6
HR = 2.393
Obs: 1
HR = 2.404
Obs: 7
HR = 2.497
Obs: 8
HR = 2.539
Obs: 9
HR = 2.549
Correlation
Time [s]
Time [s]
Time [s]

◂ **Fig. 3.8** *Left panel* Autocorrelation functions (ACFs) generated from the source counts of four soft/intermediate-state and five hard-state observations of Cygnus X-1. The width of the peak increases as the hardness ratio increases, as shown in Fig. 3.6. This is consistent with the findings of Pottschmidt et al. (2003), who reported that characteristic features of the power spectral density (PSD) shift towards higher frequencies as the spectra soften. *Centre panel* Cross-correlation functions between the source counts and the photon index (*solid green line*). I note that the width of the peak is always greater on the *left-hand side* of the y-axis than the *right-hand side* and suggest that this may be caused by the presence of at least three components in the CCFs—a positive symmetric correlation, a negative symmetric correlation and an asymmetric component. Also shown on the same plots is a reflection in the y-axis of the *left hand side* of the CCFs (*dotted purple line*). *Right panel* The difference between the *right hand side* of the CCF and the reflection of the *left hand side* in the y-axis (*green line minus purple line*). Many of these plots appear similar to the CCFs produced from the faintest hard-state observations (see Fig. 3.4), which lacked a strong positive correlation, and may therefore suggest that the asymmetric component that was clearly evident in the hard-state observations may also be present in the soft-state observations

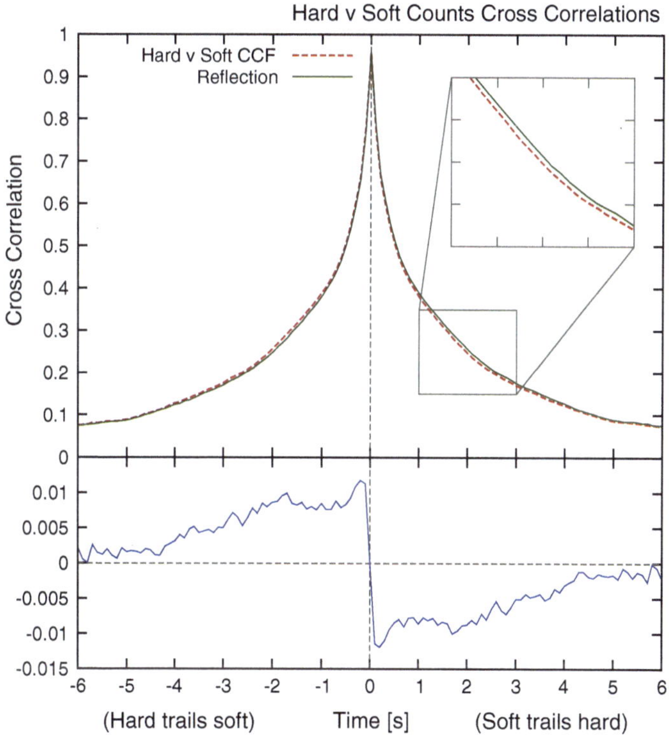


**Fig. 3.9** Cross-correlation function between the hard (5–13 keV) count rate and the soft (3–5 keV) count rate (*dashed red line*) for the combined lightcurve from 8 hard-state observations. The function is almost symmetric, but when plotted alongside its own reflection in the y-axis (*solid green line*) the slight asymmetry reveals that the hard count rate, or some component of it, slightly lags the soft count rate. The bottom panel emphasises this point by showing the difference between these two functions. In Sect. 3.10 I consider the effect that this lag may have on my measurements of Γ, and suggest that it may be the most likely explanation for the asymmetric shape of my Γ versus counts CCFs

### 3.6.2 A Lag between the Soft and Hard Counts

Further analysis of the combined 100 ms data from the eight hard-state observations reveals a small asymmetry in the CCF (Fig. 3.9, dashed red line) of hard counts (5–13 keV) versus soft counts (3–5 keV), which can only be produced if a small component of the hard count rate lags the changes in the soft count rate (referred to as hard lag). Although the peak in the CCF is almost symmetric, when plotted alongside its own reflection in the y-axis (i.e. the CCF of soft counts against hard counts, solid green line) the slight asymmetry becomes apparent, and can be further emphasised by subtracting one CCF from the other (bottom panel, solid blue line). This asymmetry in the CCF of Cygnus X-1 was first observed by Priedhorsky et al. (1979) and Nolan et al. (1981) and later re-examined in more detail by Maccarone et al. (2000) who ruled out light travel delays in the corona as an explanation.

In Sect. 3.10 I consider the effect that this lag may have on my measurements of $\Gamma$, and suggest that it may be the most likely explanation for the asymmetric shape of my $\Gamma$ versus counts CCFs.

## 3.7 A Comparison with GX 339-4

In order to investigate whether the results discussed in this chapter are unique to Cygnus X-1 or are also found in other Galactic black holes I have extended my analysis to include the low-mass X-ray binary (LMXB) system GX 339-4. In the quiescent state, this source is generally much too faint to allow spectra to be fitted on 100 ms time-scales and I am therefore restricted to using only those observations that were made at times when GX 339-4 was experiencing an outburst. However, although the source is sufficiently bright at these times, I often find that most of the photons are coming from a thermal component and not from the power-law, which can still be difficult to fit on short time-scales. I have attempted to resolve this problem by selecting datasets from times when the source was ascending towards its maximum brightness (around April and May 2002, during which time the source could be found first in a hard state, and then later a soft state), and neglecting the softer state observations which followed in subsequent months. For a discussion of GX 339-4 during this outburst see Belloni et al. (2005). The cross-correlation functions produced from these data are shown in Fig. 3.10 and reveal similar behaviour to that found in the CCFs of Cygnus X-1, namely a positive peak that becomes wider and more asymmetric as the hardness ratio increases and, in the observation with the greatest hardness ratio, the emergence of a narrow anti-correlation skewed towards positive lags.

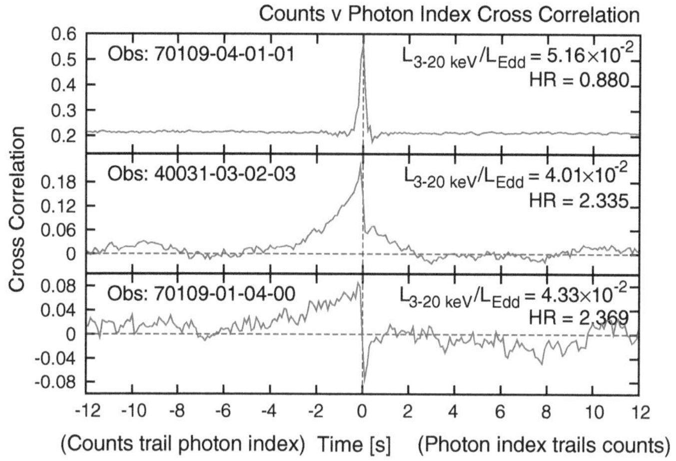


**Fig. 3.10** Cross-correlation functions between the photon index and the 3–20 keV source counts for the low-mass X-ray binary (LMXB) system GX 339-4. The panels show three observations from the *RXTE* archive, dating from April/May 2002, and ordered by hardness ratio ($L_{5-13\,\mathrm{keV}}/L_{3-5\,\mathrm{keV}}$). The results show similar behaviour to that found in the soft-state and bright hard-state observations of Cygnus X-1 (Fig. 3.4), with the shape of the CCFs reminiscent of observations 1, 8 and 10. The anti-correlation found in the faint hard state of the Cygnus X-1 CCFs is also present here in the observation with the greatest hardness ratio

## 3.8 The Effect of Noise on My Results

In the hard-state CCFs shown in Fig. 3.4, the strongest anti-correlation has a correlation coefficient of approximately $-0.2$. Such a weak correlation may be the result of noise in the data and it is therefore worth exercising a little caution. Figure 3.11 shows the autocorrelation functions (ACFs) of the source counts and $\Gamma$ for the hard-state observation 7, shown at 40 ms time resolution. The ACF of a lightcurve consisting of just white noise would have a form similar to a delta function and it is clear from the narrow peak that $\Gamma$ is heavily affected by noise. This does not mean, however, that $\Gamma$ is entirely uncorrelated with the source counts and, since a large number of spectra (between $\sim$10,000 and $\sim$100,000 for each observation) contribute to each CCF, we believe that the results are still highly significant.

In order to test this assertion we have generated a new CCF by completely randomising the photon index data while leaving the source counts unaltered (Fig. 3.12). This new CCF shows small amplitude variations that are more than an order of magnitude weaker than the peak we see in the CCF of the real data.

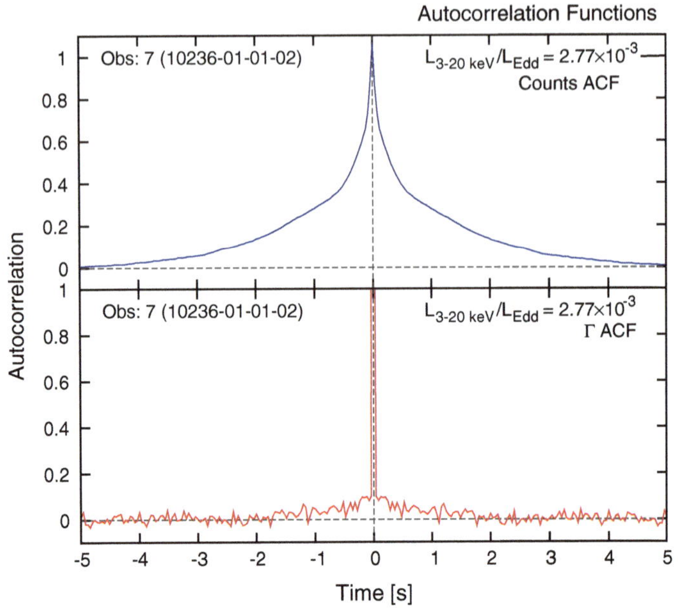


**Fig. 3.11** Autocorrelation functions (ACFs) for the hard-state observation 7, showing source counts (*top*) and Γ (*bottom*). The extremely narrow peak in the Γ ACF shows that these data are heavily affected by noise, which was already suspected from the large amount of scatter that was evident in Fig. 3.3. This does not mean, however, that Γ should be entirely uncorrelated with the source counts nor that this noise is responsible for generating the peaks found in the CCFs

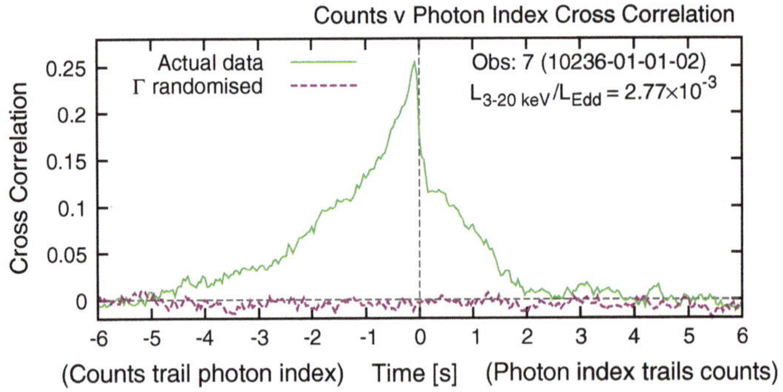


**Fig. 3.12** A comparison between the CCF for observation 7 and a new CCF, generated by completely randomising the photon index data while leaving the source counts unaltered. The new CCF shows small amplitude variations that are at least an order of magnitude weaker than the asymmetric peak we see in the real data CCF

## 3.9 Summary of the Main Observational Results

- In the hardest states of Cygnus X-1 which I have examined, the photon index, $\Gamma$, is anti-correlated with the count rate. Where a positive time lag is defined as $\Gamma$ lagging behind the count rate, there is little correlation at negative lag. However, a strong anti-correlation occurs very sharply within 16 ms of zero lag and then decreases over the following few seconds. The strongest anti-correlation has a correlation coefficient ($\sim -0.2$) which is relatively weak when compared to the strength of the positive correlations found in the soft state.
- In the softest states which I have examined, $\Gamma$ increases as the count rate increases (i.e. there is a positive correlation between the two parameters). The CCF is nearly symmetric about zero lag and the peak correlation coefficient is quite high ($> 0.75$)—see Fig. 3.5.
- When the data from the three softest of the hard/intermediate state observations are excluded (i.e. observations 1, 6 and 7), a CCF generated from the remaining hard-state data (Fig. 3.7, bottom panel) shows a weak, broad, positive correlation at negative lags and a weak, broad, negative correlation at positive lags, together with a sharp negative peak in the CCF near zero lag. The weak and broad components of the CCF are remarkably close in shape to that which was produced when I examined the difference between the hard count rate versus soft count rate CCF and its reflection about zero lag (Fig. 3.9, bottom panel).
- There is an almost monotonic transition with varying hardness ratio between the shapes of the CCF in the softest and hardest states with the CCFs in the intermediate hardness states being simply replicated as a sum of the two extreme CCFs, each added in proportion with hardness ratio.
- The full width of the CCFs in the soft state, together with twice the half-width at negative lags in the harder states, correlates well with hardness ratio, as does the full width of the ACF in all states.
- Although I have not yet investigated the behaviour of other XRBs in as much detail, I have presented some early indications that the CCFs of GX 339-4 exhibit similar features to those of Cygnus X-1, suggesting that this behaviour may be common to all black hole XRBs.

## 3.10 Discussion

The CCFs provide clear evidence for two distinct components contributing to the X-ray emission. The soft-state CCFs are the simplest to explain. They show the standard behaviour expected when an increased seed photon flux, perhaps resulting from an increase in the accretion rate through the disc, is sufficiently large that it cools the Comptonising corona and leads to both an increased X-ray flux and also a softer spectrum (Laurent and Titarchuk 2011; Qiao and Liu 2013). Thus energy is released primarily through the production of seed photons rather than in heating the corona.

Seed photon production is most efficient from the optically thick part of the accretion disc and the small width of the CCF is consistent with a change in the seed photon flux on a viscous time-scale associated with an accretion disc at small gravitational radius. The very direct link between seed photon flux and coronal cooling leads to a relatively high peak CCF value and a lag close to zero. The decreasing width of the CCF with decreasing hardness is consistent with the inward motion of the inner part of the disc with increasing accretion rate.

The hard-state CCFs are more complex. As I explained above (Sect. 3.9), they appear to consist of a broad anti-symmetric component and a narrow symmetric component (see Fig. 3.7—bottom panel). The narrow width of the symmetric component implies that the origin of the flux change and of the spectral index change for this component have to be very closely coupled and in the same physical location. Cyclo-synchrotron self-Compton emission, where the cyclo-synchrotron seed photons are scattered by their very nearby parent electrons, naturally provides this component to the CCF, and should dominate in the highest density, inner regions of the corona or RIAF where the locally produced cyclo-synchrotron photons may form the bulk of the seed photons. Enhanced cyclo-synchrotron radiation implies that more high energy electrons have been injected, i.e. energy has been put into the Comptonising electrons of the corona rather than into the seed photons.

To produce the broad anti-symmetric component I require that the soft part of the spectrum rises first, leading to a steepening of the spectrum with count rate. I then

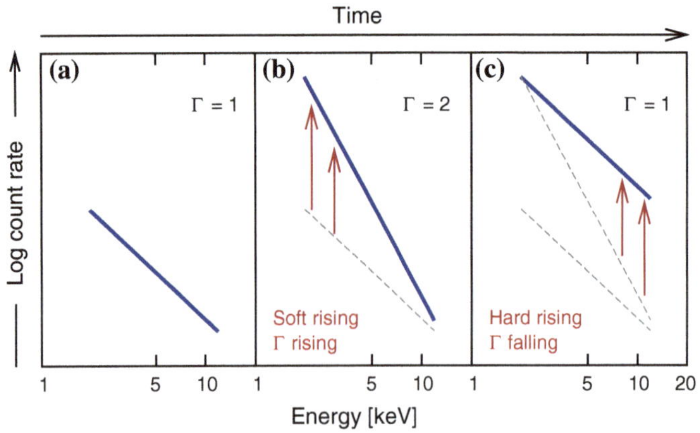


**Fig. 3.13** An illustration of an X-ray spectrum with a single power-law component (panel *a*), demonstrating how a lag between the soft count rate and the hard count rate may temporarily affect the measurement of $\Gamma$ and potentially influence the CCF of count rate versus $\Gamma$. An increase in count rate is initially observed more strongly at soft energies, which causes the spectrum to temporarily soften (panel *b*). At some later time the hard energy count rate will also start to rise and the spectrum will now begin to harden (panel *c*). Together, one might expect this softening and hardening of the spectrum to produce a positive correlation between $\Gamma$ and the total count rate at negative lags ($\Gamma$ leads count rate) and an anti-correlation at positive lags (count rate leads $\Gamma$). Observationally, the hard-state observations provide some indication that this may be the case (see Figs. 3.7, *bottom*, and 3.4, *observations* 10, 13 and 14)

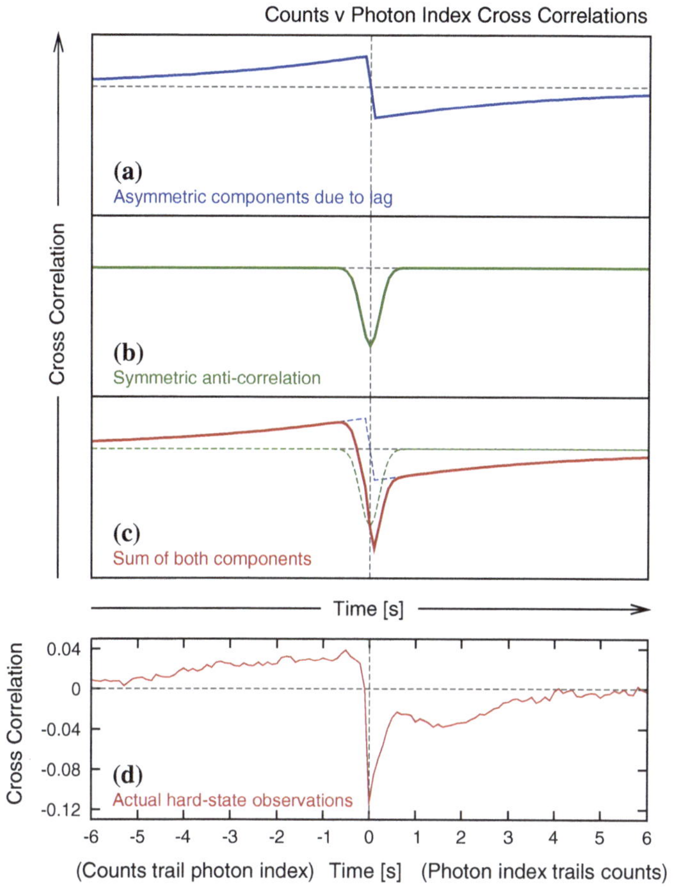


**Fig. 3.14** An illustration of a CCF between count rate and Γ, demonstrating how a lag between the soft count rate and the hard count rate may be responsible for the asymmetric shape of the hard-state CCFs. *Panel a* The highly asymmetric components introduced into the CCF by small changes in Γ as variations in the hard count rate lag those in the soft count rate (see Fig. 3.13). *Panel b* The narrow anti-correlation which I tentatively associate with Comptonisation from cyclo-synchrotron photons in the hot inner flow. *Panel c* The sum of the components in panels a and b. *Panel d* The CCF generated from eight hard-state observations, showing a positive correlation at negative lags, a negative correlation at positive lags and a narrow, asymmetric anti-correlation

require that, some time later, the count rate in the higher energy parts of the spectrum rises, leading to a hardening of the spectrum with count rate. Thus, as the count rate rises, a steepening of the spectrum (Γ is positively correlated with the count rate) precedes a hardening of the spectrum (Γ is anti-correlated with the count rate) and introduces the weak asymmetric component found in my CCFs. This scenario, which requires that some component of the hard flux lags the soft, agrees entirely with the hard versus soft band CCF (Fig. 3.9) and is illustrated in Fig. 3.13. Such spectral changes can be explained physically if there is an extended emission region with a radial gradient in the energy or temperature of the Comptonising electrons such that the higher energy electrons are mostly closer to the black hole than are the lower energy electrons. Thus the emission weighted centroid of the high energy photons would be closer to the black hole than that of the lower energy photons. An inwardly moving perturbation, perhaps in accretion rate, would then boost the low energy photon emission rate before that of the higher energy photons. This scenario is entirely consistent with what is now becoming the standard paradigm to explain most aspects of X-ray spectral variability, i.e. accretion rate perturbations, produced probably in the accretion disc (Lyubarskii 1997), propagating inwards and affecting the X-ray emission from a radially temperature-stratified emission region (e.g. Kotov et al. 2001; Churazov et al. 2001; Arévalo and Uttley 2006; Arévalo et al. 2008).

For the above propagating fluctuation model to work I require that the main X-ray production mechanism at intermediate hardness is not self-Compton scattering of locally produced cyclo-synchrotron photons, which would only produce a narrow anti-correlation. Instead, I require that, outside of the densest central region of the corona, the seed photons will be dominated by a mixture of thermal disc seed photons and a significant component of cyclo-synchrotron photons produced in more distant regions, although there will also be some locally produced cyclo-synchrotron seed photons. Thus the factor which mainly drives flux variability in the outer regions will be enhancement of the scattering electron population rather than seed photon enhancement and so lags will indeed be expected between soft and hard photons as the perturbations pass through the energy stratified scattering region.

Figure 3.14 (panel *a*) shows an illustration of how an asymmetric component, generated purely from my simple model of an energy/temperature gradient and propagation lag rather than any genuine change in $\Gamma$, might appear on the CCF of count rate versus $\Gamma$. When added to a narrower, symmetric anti-correlation (panel *b*) I find that the result is a skewed anti-correlation (panel *c*) that has at least some potential for reproducing the overall shape of the CCF generated from the eight hard-state observations (panel *d*).

## References


Arévalo, P., McHardy, I. M., & Summons, D. P. (2008). X-ray variability of the Seyfert 1 Markarian 335: Power spectrum and time lags. *Monthly Notices of the Royal Astronomical Society*, *388*, 211–218.

Arévalo, P., & Uttley, P. (2006). Investigating a fluctuating-accretion model for the spectral-timing properties of accreting black hole systems. *Monthly Notices of the Royal Astronomical Society*, *367*, 801–814.

Belloni, T., Homan, J., Casella, P., van der Klis, M., Nespoli, E., Lewin, W. H. G., et al. (2005). The evolution of the timing properties of the black-hole transient GX 339–4 during its 2002/2003 outburst. *Astronomy & Astrophysics*, *440*, 207–222.

Churazov, E., Gilfanov, M., & Revnivtsev, M. (2001). Soft state of Cygnus X-1: Stable disc and unstable corona. *Monthly Notices of the Royal Astronomical Society*, *321*, 759–766.

Done, C., & Gierliński, M. (2006). Truncated disc versus extremely broad iron line in XTE J1650–500. *Monthly Notices of the Royal Astronomical Society*, *367*, 659–668.

Edelson, R. A., & Krolik, J. H. (1988). The discrete correlation function—A new method for analyzing unevenly sampled variability data. *Astrophysical Journal*, *333*, 646–659.

Esin, A. A., McClintock, J. E., & Narayan, R. (1997). Advection-dominated accretion and the spectral states of black hole X-Ray binaries: Application to nova MUSCAE 1991. *Astrophysical Journal*, *489*, 865.

Gu, M., & Cao, X. (2009). The anticorrelation between the hard X-ray photon index and the Eddington ratio in low-luminosity active galactic nuclei. *Monthly Notices of the Royal Astronomical Society*, *399*, 349–356.

Ishibashi, W., & Courvoisier, T. J.-L. (2012). The physical origin of the X-ray power spectral density break timescale in accreting black holes. *Astronomy & Astrophysics*, *540*, L2.

Kotov, O., Churazov, E., & Gilfanov, M. (2001). On the X-ray time-lags in the black hole candidates. *Monthly Notices of the Royal Astronomical Society*, *327*, 799–807.

Laurent, P., & Titarchuk, L. (2011). Spectral index as a function of mass accretion rate in black hole sources: Monte Carlo simulations and an analytical description. *Astrophysical Journal*, *727*, 34.

Lyubarskii, Y. E. (1997). Flicker noise in accretion discs. *Monthly Notices of the Royal Astronomical Society*, *292*, 679.

Maccarone, T. J., & Coppi, P. S. (2002). Short-time-scale correlations between line and continuum fluxes in Cygnus X-1. *Monthly Notices of the Royal Astronomical Society*, *335*, 465–472.

Maccarone, T. J., Coppi, P. S., & Poutanen, J. (2000). Time domain analysis of variability in Cygnus X-1: Constraints on the emission models. *Astrophysical Journal Letters*, *537*, L107–L110.

Magdziarz, P., & Zdziarski, A. A. (1995). Angle-dependent Compton reflection of X-rays and gamma-rays. *Monthly Notices of the Royal Astronomical Society*, *273*, 837–848.

Miniutti, G., & Fabian, A. C. (2004). A light bending model for the X-ray temporal and spectral properties of accreting black holes. *Monthly Notices of the Royal Astronomical Society*, *349*, 1435–1448.

Narayan, R., & McClintock, J. E. (2008). Advection-dominated accretion and the black hole event horizon. *New Astronomy Review*, *51*, 733–751.

Niedźwiecki, A., Xie, F.-G., & Zdziarski, A. A. (2012). General relativistic model of hot accretion flows with global Compton cooling. *Monthly Notices of the Royal Astronomical Society*, *420*, 1195–1206.

Nolan, P. L., Gruber, D. E., Matteson, J. L., Peterson, L. E., Rothschild, R. E., Doty, J. P., et al. (1981). Rapid variability of 10–140 keV X-rays from Cygnus X-1. *Astrophysical Journal*, *246*, 494–501.

Nowak, M. A., Wilms, J., & Dove, J. B. (2002). Coronal-temporal correlations in GX 339–4: Hysteresis, possible reflection changes and implications for advection-dominated accretion flows. *Monthly Notices of the Royal Astronomical Society*, *332*, 856–878.

Orosz, J. A., McClintock, J. E., Aufdenberg, J. P., Remillard, R. A., Reid, M. J., Narayan, R., et al. (2011). The mass of the black hole in Cygnus X-1. *Astrophysical Journal*, *742*, 84.

Pottschmidt, K., Wilms, J., Nowak, M. A., Pooley, G. G., Gleissner, T., Heindl, W. A., et al. (2003). Long term variability of Cygnus X-1. I. X-ray spectral-temporal correlations in the hard state. *Astronomy & Astrophysics*, *407*, 1039–1058.

Priedhorsky, W., Garmire, G. P., Rothschild, R., Boldt, E., Serlemitsos, P., & Holt, S. (1979). Extended-bandwidth X-ray observations of Cygnus X-1. *Astrophysical Journal*, *233*, 350–363.

Qiao, E., Liu, B. F. (2010). The spectral features of disk and corona with mass evaporation in the low/hard state. *Publications of the Astronomical Society of Japan*, 62:661.

Qiao, E., & Liu, B. F. (2013). A model for the correlation of hard X-Ray index with Eddington ratio in black hole X-Ray binaries. *Astrophysical Journal*, *764*, 2.

Reid, M. J., McClintock, J. E., Narayan, R., Gou, L., Remillard, R. A., & Orosz, J. A. (2011). The trigonometric parallax of Cygnus X-1. *Astrophysical Journal*, *742*, 83.

Skipper, C. J., McHardy, I. M., & Maccarone, T. J. (2013). Very fast X-ray spectral variability in Cygnus X-1: Origin of the hard- and soft-state emission components. *Monthly Notices of the Royal Astronomical Society*, *434*, 574–584.

Sobolewska, M. A., Papadakis, I. E., Done, C., & Malzac, J. (2011). Evidence for a change in the X-ray radiation mechanism in the hard state of galactic black holes. *Monthly Notices of the Royal Astronomical Society*, *417*, 280–288.
Sunyaev, R. A., & Titarchuk, L. G. (1980). Comptonization of X-rays in plasma clouds—Typical radiation spectra. *Astronomy & Astrophysics*, *86*, 121–138.
Wu, Q., & Gu, M. (2008). The X-Ray spectral evolution in X-Ray binaries and its application to constrain the black hole mass of ultraluminous X-Ray sources. *Astrophysical Journal*, *682*, 212–217.
Wu, Y. X., Belloni, T. M., & Stella, L. (2010). Fast spectral variability from Cygnus X-1. *Monthly Notices of the Royal Astronomical Society*, *408*, 2413–2425.
Zdziarski, A. A., Lubiński, P., & Smith, D. A. (1999). Correlation between Compton reflection and X-ray slope in Seyferts and X-ray binaries. *Monthly Notices of the Royal Astronomical Society*, *303*, L11–L15.
Zdziarski, A. A., Poutanen, J., Paciesas, W. S., & Wen, L. (2002). Understanding the long-term spectral variability of Cygnus X-1 with burst and transient source experiment and all-sky monitor observations. *Astrophysical Journal*, *578*, 357–373.

# Chapter 4
# The $\dot{m}_{\mathrm{E}}$–Γ Relationship in XRBs and AGN

*Not only is the Universe stranger than we think, it is stranger than we can think.*

Werner Heisenberg (1901–1976)

## 4.1 Synopsis

In this chapter I continue my strategy of very high time resolution spectral fitting of XRBs, and extend this analysis to include a study of AGN on comparable (mass-scaled) time-scales. As well as cross-correlating count rate with Γ for the XRBs, I also demonstrate that the spectra extracted from each 100 ms time-block can be summed together with other spectra of similar count rates to improve the signal-to noise ratio and allow a direct examination of the distribution of Γ with $\dot{m}_{\mathrm{E}}$. This technique is then applied to AGN using spectra extracted from long-term *RXTE* monitoring observations instead of the 100 ms data sections that were used for the XRBs.

## 4.2 Introduction

The relationship between the bolometric luminosity of accreting black holes and the observed X-ray photon index (Γ) has already been extensively discussed in the introduction, as well as in Chap. 3. Here, I will briefly recap upon those elements of this relationship that are most relevant to this section, and extend the discussion to include a comparison of the behaviour observed in AGN and XRBs.

The power-law component found in the X-ray spectra of AGN and XRBs (which takes the form $N_\gamma \propto E^{-\Gamma}$, where $N_\gamma$ is the photon number density, $E$ is the energy and Γ is the photon index) is the result of Comptonisation of lower energy seed



C. Skipper, *Fast Spectral Variability in the X-ray Emission of Accreting Black Holes*, Springer Theses, DOI 10.1007/978-3-319-09587-5_4

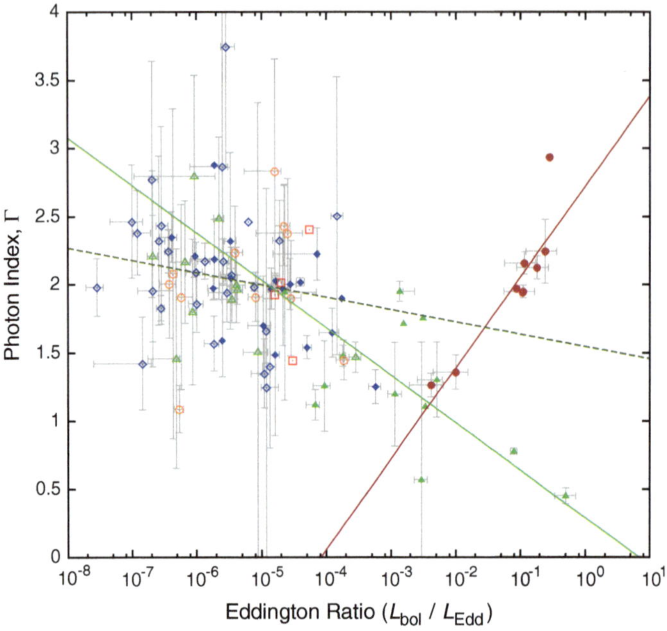


**Fig. 4.1** Accretion rate versus photon index for a sample of LLAGN and quasars. *Green triangles* represent Seyferts, *blue diamonds* represent LINERs, *red squares* represent H II regions, *open orange circles* represent transition regions and *filled red circles* represent PG quasars. For the Seyferts and LINERs, *filled symbols* represent spectral types 1.0–1.9 and *open symbols* represent type 2.0. The *solid green line* ($\Gamma = -0.35 \log(L_{bol}/L_{Edd}) + 0.30$) is a fit to the LLAGN sources, the *solid red line* ($\Gamma = 0.67 \log(L_{bol}/L_{Edd}) + 2.73$) is a fit to the quasars and the *dashed green line* ($\Gamma = -0.09 \log(L_{bol}/L_{Edd}) + 1.55$) is the best fit found by Gu and Cao (2009) for a similar sample of LLAGN

photons by high energy electrons in either a central corona or the base of a jet. Observationally, Γ has been found to be positively correlated with the source flux above a critical accretion rate ($\dot{m}_{crit}$), typically 0.5–1 % of the Eddington limit, and anti-correlated at lower accretion rates (Constantin et al. 2009; Gu and Cao 2009; Shemmer et al. 2006; Wu and Gu 2008; Sobolewska et al. 2011; Zdziarski et al. 2002). This change in the $\dot{m}_E$–Γ relationship at $\dot{m}_{crit}$ has been observed in both AGN and the hard state of XRBs, and, in the latter, it has been observed over both long (days to years) and short (sub-second) time-scales.

In Fig. 4.1 I show the relationship between $\dot{m}_E$ (or $L_{bol}/L_{Edd}$) and Γ for the same sample of AGN that I discussed extensively in Chap. 2. These data are all based upon 0.3–10 keV spectral fits to archival *Chandra* observations of the 486 galaxies in the Palomar sample ( Ho et al. 1997) and, in addition, a small number of Palomar-Green (PG) quasars. The plot shows a strong anti-correlation amongst the Palomar sources, with Γ typically between 2 and 2.5 at accretion rates below $\sim 10^{-6}$. In addition, the strength of this correlation is likely to improve if only type 1.0–1.9 sources are considered (filled symbols), which indicates that a reduction in contamination from non-AGN sources would enhance the clarity of this "harder-when-brighter" behaviour. In contrast to the LLAGN, the quasars exhibit a positive correlation between $\dot{m}_E$ and $\Gamma$ (which was also observed in a similar sample of quasars by Shemmer et al. 2006) that, once fitted with a straight line, intersects the fit to the LLAGN at an accretion rate of $\sim 3.5 \times 10^{-3}$. This value is consistent with that at which hard-state XRBs switch from an anti-correlation to a positive correlation, and is lower than the accretion rate at which XRBs make the transition from the hard to the soft state (around a few percent of the Eddington limit—see Maccarone 2003; Gierliński and Newton 2006).

The anti-correlation found in my LLAGN *Chandra* data (Fig. 4.1, solid green line) is somewhat steeper than that reported by Gu and Cao (2009) (I find a slope of $-0.35$, compared to their slope of $-0.09$). This discrepancy is possibly due to a slight bias introduced by the presence in my data of several Compton-thick AGN at high accretion rates, but may also be the result of greater scatter in their data, which was collected from multiple literature sources (the authors make a similar point in their discussion).

Although large snapshot surveys of AGN present a simple and effective method of studying the $\dot{m}_E$–$\Gamma$ relationship, it would also be useful to examine how individual sources evolve on this plot. It has been shown (e.g. Ponti et al. 2006) that some AGN exhibit correlations between $\Gamma$ and the source flux on time-scales of only a few hours, and their evolution on the $\dot{m}_E$–$\Gamma$ plot can therefore potentially be examined using a single *XMM* or *Chandra* observation. However, in anticipation of large uncertainties in $\Gamma$, I instead opt to study individual AGN over much longer time-scales (weeks or months) and thereby hope to observe the changes in $\Gamma$ over a much larger variation in flux. In addition, this strategy has the further advantage of providing a potentially greater number of data points and a greater number of counts within each spectrum.

In order to compare the spectral variability of AGN with that of XRBs, it is necessary to consider the black hole mass ($M_{BH}$) of the source. McHardy et al. (2006) showed that variability time-scales scale with $M_{BH}$ and inversely with $\dot{m}_E$, which means that variability behaviour observed in AGN on time-scales of weeks should, if present in XRBs, be found by studying variability on time-scales of seconds. For this reason, I choose to examine the spectral variability of AGN by using multiple observations from long monitoring campaigns, and examine the spectral variability of XRBs by monitoring the rapid variations in $\Gamma$ and flux on millisecond time-scales within individual observations.

## 4.3 Observations and Data Reduction

### *4.3.1 The Sample*

During its lifetime, the *RXTE* satellite undertook many long-term monitoring campaigns of AGN and, in this chapter, I choose to focus on five sources that have accretion rates below a few percent of the Eddington limit. Four of these sources (NGC 1052, NGC 3516, NGC 3998 and NGC 5548) are part of the Palomar sample, which was discussed extensively in Chap. 2. The X-ray spectrum of the LINER-1.9 galaxy NGC 1052 has been the subject of a number of papers (e.g. Weaver et al. 1999; Guainazzi and Antonelli 1999), all of which remark upon its unusual flatness in the 2–10 keV range. In Chap. 2 I included NGC 1052 as a potential Compton-thick candidate, but noted that very little evidence has been found to support this inclusion beyond its unusually flat spectrum; in addition, the lack of a strong iron line in this source (Guainazzi and Antonelli 1999) is not consistent with other known Compton-thick candidates. The other three Palomar sources included in the sample (NGC 3516, NGC 3998 and NGC 5548) are all Compton-thin type-1 Seyferts or LINERs, chosen primarily due to the relative simplicity of their spectra, which show no evidence of a significant soft excess or similar features which could complicate the fitting process. The fifth source I have chosen to include in my sample is the Seyfert/LINER hybrid NGC 7213, which is already known to demonstrate an anti-correlation between $\dot{m}_E$ and Γ (Emmanoulopoulos et al. 2012) over long time-scales.

The choice of X-ray binary systems suitable for inclusion in my sample is heavily restricted by the available source flux. If the AGN observations are typically sampled at a frequency of one observation every four days (or $\sim 3.5 \times 10^5$ s), then this implies a spectral timing resolution for XRBs of ~100 ms (assuming a typical AGN has a black hole mass of $\sim 5 \times 10^7\ M_\odot$ and a typical XRB a mass of $\sim 10\ M_\odot$). In order to fit spectra on time-scales as short as this, I require a count rate of at least several hundred counts per second and, furthermore, a source in which the X-ray spectra are not dominated by the disc black body component. However, it is also desirable to include sources that cover as large a range of accretion rates as possible in order to examine their behaviour either side of the $\dot{m}_{crit}$ boundary and to enable a comparison with the AGN, which may cover several orders of magnitude in accretion rate. In consideration of the above restrictions, the sample was defined to be the high-mass binary system Cygnus X-1, the low-mass binary systems GX 339-4 and XTE J1118+480.

A summary of some of the key properties of the sample is provided in Table 4.1.

### *4.3.2 Data Reduction*

All observations used in this chapter have been selected and retrieved from the *RXTE* archive. In total, 115 observation of NGC 1052 were retrieved, along with 627 observations of NGC 3516, 376 observations of NGC 3998, 1,055 observations of NGC 5548 and 238 observations of NGC 7213. In addition, *all* observations of Cygnus X-1, GX 339-4 and XTE J1118+480 were retrieved.

For all observations good time intervals (GTIs) were generated based upon the spacecraft elevation angle being greater than 10°, the pointing offset being less than 0.02° and the time since the last south Atlantic anomaly (SAA) passage being at least 30 min. Next, synthetic background data based upon the faint (for the AGN) or bright (for the XRBs) PCA background models were generated using the script RUNPCABACKEST and suitable response files created using the PCARSP tool. Spectra were then extracted from both the standard-2 data and background data for PCU2

**Table 4.1** A summary of the sample

| Source | AGN/XRB | Distance (kpc) | $M_{BH}$ ($M_{\odot}$) | RA [e] | Dec [e] |
|---|---|---|---|---|---|
| NGC 1052 | AGN | 17 800 [j] | $1.80 \times 10^8$ [g] | 02h41m04.8s | −08d15m20.8s |
| NGC 3516 | AGN | 38 900 [j] | $9.03 \times 10^7$ [g] | 11h06m47.5s | +72d34m06.7s |
| NGC 3998 | AGN | 21 600 [j] | $7.32 \times 10^8$ [g] | 11h57m56.1s | +55d27m12.9s |
| NGC 5548 | AGN | 67 000 [j] | $6.09 \times 10^8$ [g] | 14h17m59.5s | +25d08m12.5s |
| NGC 7213 | AGN | 22 120 [k] | $9.77 \times 10^7$ [f] | 22h09m16.3s | −47d10m00.0s |
| Cygnus X-1 | XRB | $1.86^{+0.12}_{-0.11}$ [a] | 14.8±1.0 [b] | 19h58m21.7s | +35d12m05.8s |
| GX 339-4 | XRB | >7.00 [c] | 5.8 [d] | 17h02m49.4s | −48d47m22.8s |
| XTE J1118+480 | XRB | 1.81 [h] | 8.53 [i] | 11h18m10.8s | +48d02m12.6s |

[a] Reid et al. (2011)
[b] Orosz et al. (2011)
[c] Zdziarski et al. (2004)
[d] Hynes et al. (2003)
[e] From SIMBAD Astronomical Database
[f] Woo and Urry (2002)
[g] Based upon the relation between black hole mass and velocity dispersion from Tremaine et al. 2002, with the velocity dispersion data obtained from Ho et al. 2009
[h] Average of distances estimated by Wagner et al. (2001) (1.9 ± 0.4 kpc), McClintock et al. (2001) (1.8 ± 0.6 kpc) and Gelino et al. (2006) (1.72 ± 0.1 kpc)
[i] Gelino et al. (2006)
[j] Ho et al. (1997)
[k] Emmanoulopoulos et al. (2012)

only (a restriction designed to improve the quality of the spectra, especially those from the faintest sources).

For the XRBs, response files were also generated for the binned array modes (where available) or, alternatively, the event mode data. The observations selected from the archive all provide time resolution of 100 ms or better, and are therefore able to support the study of XRBs on (mass-scaled) time-scales that are comparable to those we can study in my AGN observations. The binned array modes are the simplest to reduce, and are therefore preferred, but offer less flexibility than the event modes (for instance, binned array mode data generally contains data from *all* PCUs binned into a single column (XeCnt), and therefore prevent any possibility of using only PCU2 data). Once response files had been generated, spectra were then extracted from each 100 ms section of the data (as described in Sect. 3.4).

For the AGN, high time resolution is not required and therefore only the standard-2 datasets have been reduced. This mode offers 16 s time-resolution and at least 40 channels to cover the 3–20 keV energy range.

### 4.3.3 Spectral Fitting

In order to determine suitable models for fitting the 100 ms XRB spectra, we first fit the standard-2 data for each full observation using XSPEC version 12.8.0. The higher spectral resolution of these data make them particularly suitable for fine tuning parameters such as iron line width and equivalent width, absorption column density and reflection scaling factor. All XRB observations were initially fitted over the 3–20 keV energy range with a model consisting of absorption (PHABS), a power-law and reflection component (PEXRAV; Magdziarz and Zdziarski 1995), a Gaussian representing the iron K $\alpha$ fluorescence line at ~6.4 keV (using the GABS component, with the line depth set to a negative value to model an emission line) and, if required, a disc black body component (DISKBB). The inclination angle of the PEXRAV component was fixed to 30° for Cygnus X-1 and 53° for GX 339-4 and XTE J1118+480.

Once suitable models had been established for each observation, the iron line width and equivalent width, absorption, reflection scaling factor and disc component were fixed, and all the 100 ms binned array mode spectra were fitted in XSPEC. From these fits I record the photon index, count rate and absorption corrected flux, along with their uncertainties. Further discussion of the validity of fixing various spectral components can be found in Sect. 3.4.

For the AGN observations, only the standard-2 data have been fitted. This was performed in XSPEC using a model consisting of an absorbed power-law ($N_H$ was set to $3.07 \times 10^{20}$ cm$^{-2}$ for NGC 1052, $3.05 \times 10^{20}$ cm$^{-2}$ for NGC 3516, and $1.22 \times 10^{20}$ cm$^{-2}$ for NGC 3998, $1.69 \times 10^{20}$ cm$^{-2}$ for NGC 5548 and $1.06 \times 10^{20}$ cm$^{-2}$ for NGC 7213). In addition, NGC 3516 and NGC 7213 each required an iron line at ~6.4 keV (modelled using the GABS component) with one sigma widths 0.76 and 0.14 keV respectively and equivalent widths 0.61 and 0.15 keV respectively.

## 4.4 The $\dot{m}_E$–Γ Relationship in X-ray Binary Systems

### *4.4.1 Short Time-Scale Data*

Previous studies of the $\dot{m}_E$–Γ relationship in hard-state XRBs on very short time-scales are fairly uncommon, but have shown (e.g. Wu et al. 2010) the correlation to be heavily affected by Poisson noise. This issue is illustrated in Fig. 3.3, in which the $\dot{m}_E$–Γ correlations in both the hard and soft-state data have been completely buried in noise. In order, therefore, to investigate this relationship in my 100 ms data I chose in Chap. 3 to cross-correlate Γ with the source count rate. This technique allowed the Poisson noise (which is random) to cancel and successfully resolved the shape of the peak in the CCF (see Fig. 3.4). However, although CCFs are a useful tool for investigating XRBs, they cannot be similarly used to study the $\dot{m}_E$–Γ relationship in AGN. The much slower variability from AGN, combined with their relative faintness, means that spectra from these sources must be extracted from complete observations that are spaced several days apart, and few (if any) AGN have a sufficient number of observations in the archives to permit this kind of analysis.

Instead, in this chapter I employ another method which allows the behaviour of XRBs and AGN to be directly compared. For the XRBs, the 100 ms PCA spectra are sorted into order of count rate, grouped together into groups of 600 (unless otherwise stated) and added together using the ADDSPEC tool. In doing this I have chosen to

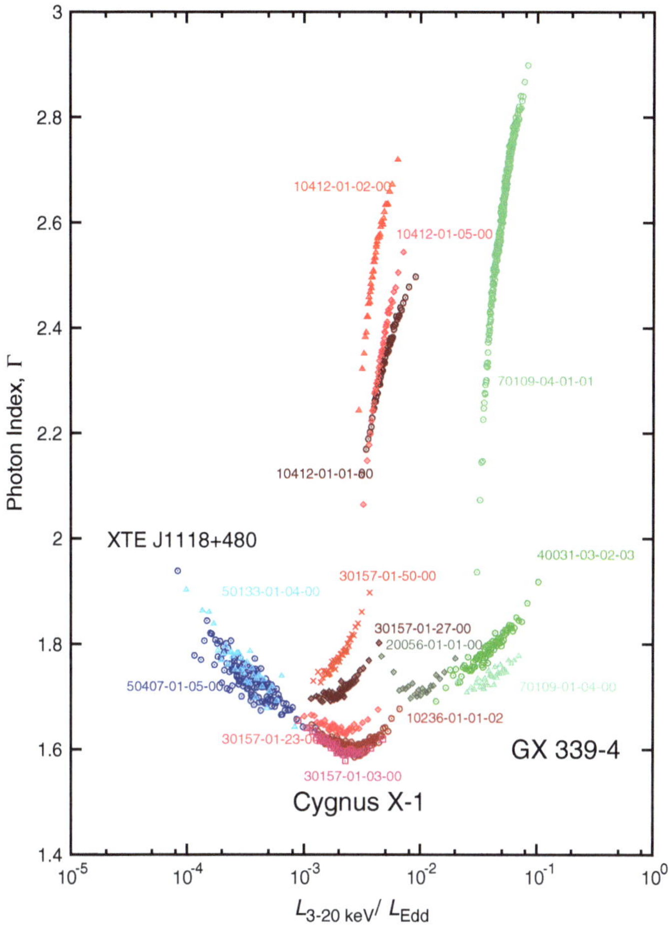


**Fig. 4.2** Accretion rate (using $L_{3-20\,\mathrm{keV}}/L_{\mathrm{Edd}}$ as a proxy) versus photon index for the X-ray binary systems XTE J1118+480, Cygnus X-1 and GX 339-4. The hard and soft states are easily distinguishable, with the latter occupying the *top* half of the plot ($\Gamma > 2$). The hard-state data show an anti-correlation at low accretion rates and a positive correlation at high accretion rates

group spectra only with others from the same observation, and all other spectra are processed entirely separately. For the XRBs, it is not necessary to combine the RMF files and background spectra when running ADDSPEC because these files have been generated based upon the entire observation, and are unlikely to change significantly on 100 ms time-scales.

Figure 4.2 shows a plot of $L_{3-20\,\mathrm{keV}}/L_{\mathrm{Edd}}$ versus Γ for eight observations of Cygnus X-1, four observations of GX 339-4 and two observations of XTE J1118+480. The soft states, in which $\Gamma > 2$, are clearly distinct from the hard and intermediate states and show very strong and steep positive correlations. At low accretion rates the hard-state observations show an anti-correlation in XTE J1118+480, and at high accretion rates a positive correlation in GX 339-4. Perhaps more interesting is the case of Cygnus X-1, which is intermediate in accretion rate between the other two sources and clearly shows both an anti-correlation and a positive correlation in the two hard-state observations. The intermediate states (30157-01-27-00 and 30157-01-50-00) show a steeper correlation and the anti-correlation that is evident in the hard state has disappeared even though these observations extend over roughly the same range of accretion rates. This behaviour in Cygnus X-1 is consistent with the count rate versus Γ CCFs described in Chap. 3, in which many of the hard-state observations showed a narrow anti-correlated component and a broader correlated component. In my discussion of those results I suggested that the broad, positive correlation represented Comptonisation of accretion disc black body photons and the narrow anti-correlation represented Comptonisation of cyclo-synchrotron emission in the hot, inner regions of the accretion flow.

The distribution of data points from the hard-state observations in Fig. 4.2 shows a remarkable continuity between sources. For instance, the position and correlation slope of XTE J1118+480 data points makes these data appear to be a continuation of the anti-correlation observed in Cygnus X-1; similarly, many of the data points representing GX 339-4 could be described as a continuation of the positive correlation observed in Cygnus X-1, with the exception of those from observation 20056-01-01-00. This latter observation, dating from 1997 April 5, appears to suggest the presence of an anti-correlation in GX 339-4 at a much higher accretion rate than the anti-correlation observed in Cygnus X-1. However, the faintness of this observation, combined with the relatively short observation time of 2.37 ks and lack of corroborative support from other observations, somewhat reduces the certainty of this result.

The assertion that the anti-correlation in Cygnus X-1 is associated with the narrow component found in the CCFs of Chap. 3 is supported by data shown in Fig. 4.3. Here, I have extracted spectra from the binned PCA data on 100 ms, 1.5 and 5 s time-scales, and added these spectra in groups of 600, 60 and 20 respectively. The plot shows that, although the positive correlation is clearly evident on all time-scales, the anti-correlation weakens in the lower time-resolution data. This result is consistent with the "two-zone" accretion model, in which seed photons from the accretion disc are responsible for the slower variability from the outer regions of the corona and cyclo-synchrotron seed photons from the hot, inner accretion flow dominate the more rapid variability.

The presence of both an anti-correlation and a positive correlation (indicating coronal heating and cooling, respectively, in response to increased flux) in the 100 ms data demonstrates the capability of the source of making this transition within a time-scale defined by the length of the observation (20 h in this case) and, in all likelihood, considerably faster than that. If it can be established that a change in the $\dot{m}_E$–Γ relationship can occur on very short time-scales then this may suggest that the cause of the transition is not a major change in the geometry of the accretion flow, but instead the data may be explained as the result of inwardly propagating accretion rate perturbations causing fluctuations in seed photon production in different parts of the accretion flow.

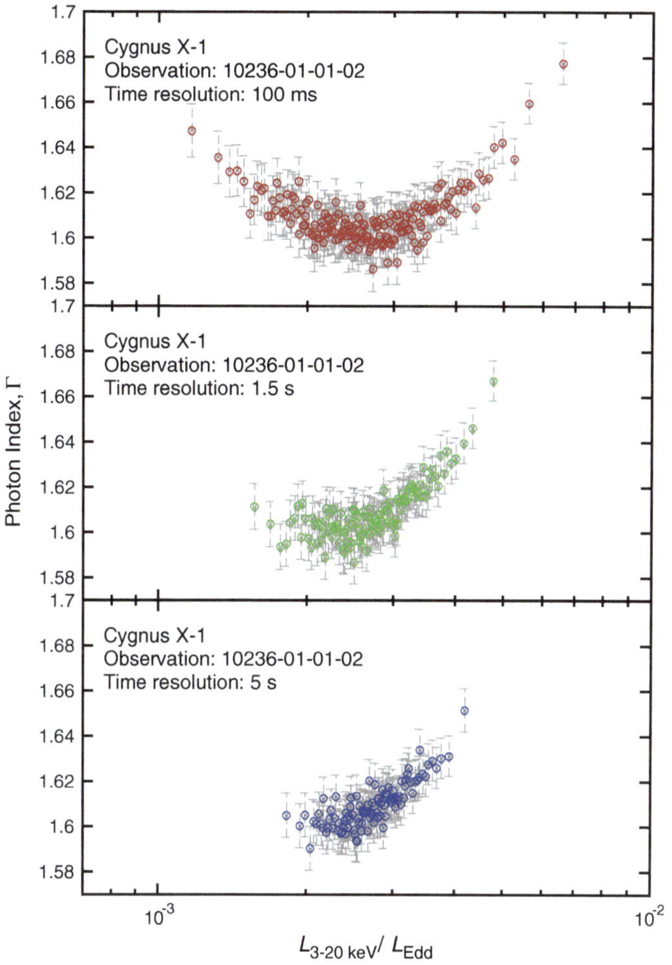


**Fig. 4.3** Accretion rate (using $L_{3-20\,\mathrm{keV}}/L_{\mathrm{Edd}}$ as a proxy) versus photon index for the X-ray binary system Cygnus X-1, observation 10236-01-01-02 (17th December 1996). *Top* Spectra were extracted from the binned PCA data at 100 ms time-resolution, sorted by count rate and then added together in groups of 600 (using the ADDSPEC tool). The summed spectra were then fitted in XSPEC. This plot demonstrates how an anti-correlation at low $\dot{m}_E$ and a positive correlation at high $\dot{m}_E$ can be found within a single 72 ks observation. *Centre* Plot generated from the same data as above, except the spectra were originally extracted at 1.5 s time-resolution instead of 100 ms. The spectra were then added together in groups of 60. *Bottom* Spectra originally extracted at 5 s time-resolution and added together in groups of 20. The anti-correlation found in the high time-resolution data becomes weaker at lower time-resolution, suggesting that the harder-when-brighter behaviour we see in Cygnus X-1 exists only within the most rapid variability

Figure 4.4 provides some further evidence in support of the idea that the switch in the $\dot{m}_E$–$\Gamma$ correlation can be extremely rapid. In this figure I have divided all the data into those where the count rate is greater than 4,000 counts per second (marked in red) and those where the count rate is less than this (marked in blue).

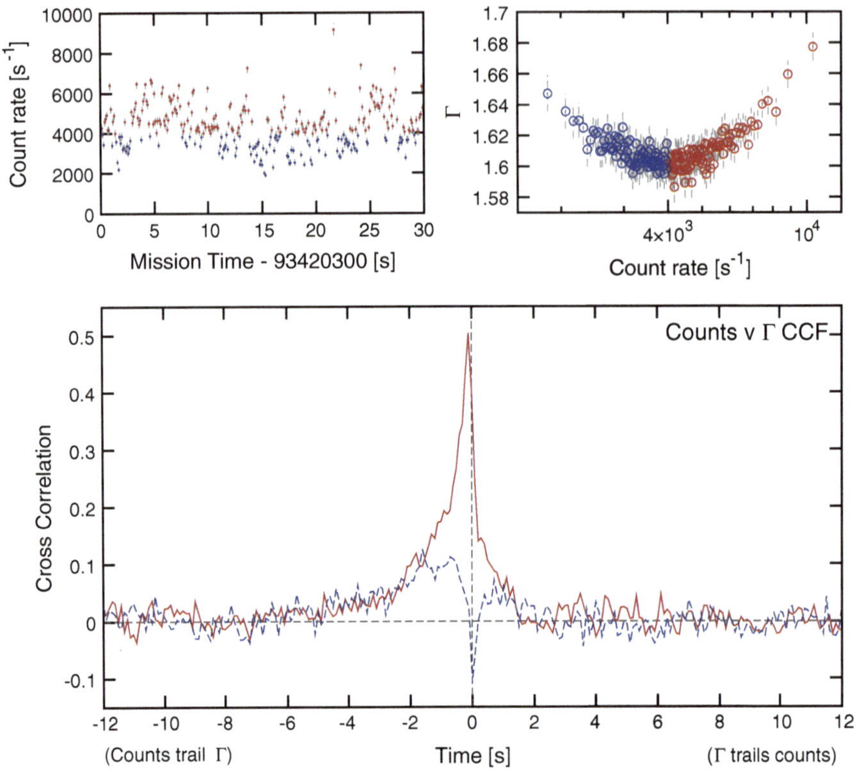


**Fig. 4.4** *Top left* First 30 s of the lightcurve of Cygnus X-1 observation 10236-01-01-02, with the data divided into those with count rate greater than (*red points*) and less than (*blue points*) 4,000 counts per second. *Top right* Plot of count rate versus Γ for the same data, with spectra sorted by count rate and added together in groups of 600. The data have again been divided by count rate. *Bottom* CCF of count rate versus Γ for the same data. As expected, the high count rate data (*solid red line*) shows a positive correlation between the two properties and the low count rate data (*dashed blue line*) shows an anti-correlation. The anti-correlation is considerably narrower and weaker than the positive correlation

A value of 4,000 cts $s^{-1}$ was chosen so that a plot of count rate versus Γ (top-right-hand corner of the figure) is divided neatly between the anti-correlation at low count rates and the positive correlation at high count rates. The source lightcurve (part of which is shown in the top-left-hand corner of the figure) shows that the source count rate repeatedly crosses this 4,000 cts $s^{-1}$ threshold on sub-second time-scales, and therefore we require a model that can explain the changes in the $\dot{m}_E$–Γ correlation on similar time-scales. The bottom panel of the figure shows the CCF of count rate versus Γ, with separate CCFs generated from the high and low count rate data (in each case there were no restrictions placed upon Γ). As expected, the high count rate CCF (red line) shows a strong positive correlation and low count rate CCF (blue line) shows a weaker anti-correlation. However, the plot also confirms that the anti-correlation found in the fainter observations is much narrower than the positive correlation, which is consistent with the disappearance of the anti-correlation in Fig. 4.3 when low time-resolution data were analysed. The low count rate CCF also shows positive peaks at around $\pm 1$–2 s, which suggests that the broader positive peak found in the high count rate CCF persists at lower count rates. The width of this positive peak is exactly the same in both the high and low count rate CCFs (about 4–5 s at the base), indicating that the inner edge of the accretion disc is fixed in truncation radius on the short time-scales examined here.

### 4.4.2 Long Time-Scale Data

The investigation of the $\dot{m}_E$–$\Gamma$ relationship over different time-scales (Fig. 4.3) can be extended to much longer time-scales of days or weeks. In order to do this it is necessary to reduce the *XTE* standard-2 data from as many observations as possible, and then group and add these spectra in the same way that was described above for the binned PCA spectra. Figure 4.5 shows the same data that was presented in Fig. 4.2 (grey points) and, in addition, data extracted from the summed standard-2 spectra of all three sources. The observations from which these spectra were extracted are typically found in the archive with a gap between observations of between one and four days, and several hundred observations were retrieved for each source. The plot shows that, on long time-scales, XTE J1118+480 shows remarkably similar behaviour to that which was observed in the 100 ms PCA data, suggesting that the temperature/luminosity relation in the advection dominated accretion flow of this source is independent of the time-scale. In contrast, Cygnus X-1 shows only a very steep correlation that is consistent in slope with that observed in the 100 ms PCA data from the soft-state observations. Unfortunately, there is no archive data that catches Cygnus X-1 in a lower flux state so the presence of an anti-correlation on long time-scales cannot be verified in this source. In GX 339-4 we find a positive correlation at high $\dot{m}_E$ that is consistent with the 100 ms PCA data. In this plot I only show the data up to the time when the source enters the intermediate state (2010 January 12 to 2010 April 09), as the uncertainty on $\Gamma$ was found to be too large after this point. In addition, the plot shows data from a period in which GX 339-4 was leaving the soft state and returning to the hard state (2003 March 29 to 2003 April 27), and here we find a much steeper distribution of $\Gamma$ at a lower $\dot{m}_E$ that is more alike to the long-term data from Cygnus X-1. The different $\dot{m}_E$–$\Gamma$ distributions of GX 339-4 when entering and leaving the soft state are reflective of the hysteresis curve found in the hardness-luminosity diagram of this source (see Sect. 4.4.4).

GX 339-4 makes the hard state to hard-intermediate state transition at a much higher accretion rate than Cygnus X-1, such that the accretion rate of the transition ($\dot{m}_{H\rightarrow HI}$) is at least an order of magnitude greater than $\dot{m}_{crit}$. Consequently, GX 339-4 displays a positive correlation in the long-term hard-state data that is not present in the Cygnus X-1 data. A further examination of the hard-state GX 339-4 data reveals an increase in the equivalent width of the iron line with count rate (see Figs. 4.6 and 4.7 in both the long- and short-term observations. This is somewhat unexpected, and is inconsistent with the results found for Cygnus X-1 in Chap. 3 and for GX 339-4 by Nowak et al. (2002).

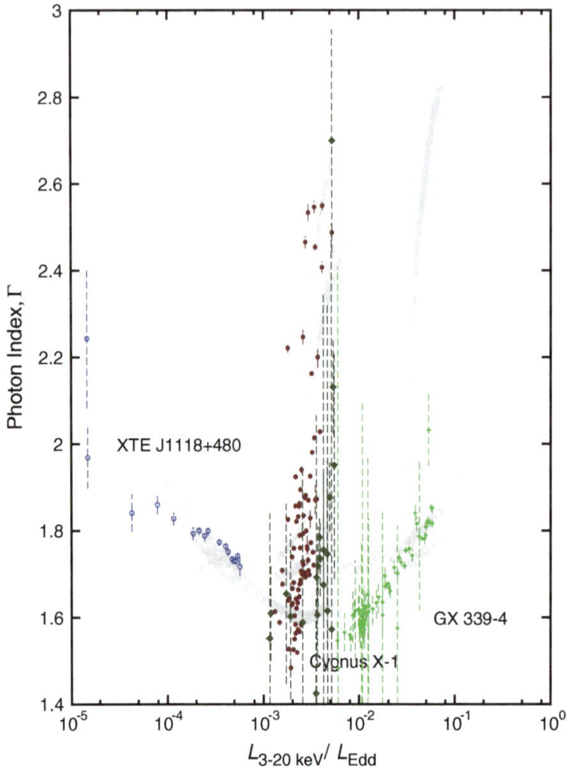


**Fig. 4.5** Accretion rate (using $L_{3-20\,\mathrm{keV}}/L_{\mathrm{Edd}}$ as a proxy) versus photon index for Cygnus X-1 (*filled red circles*), XTE J1118 + 480 (*open blue circles*) and GX 339-4 (*filled green diamonds*) over long time-scales (months and years); the *grey points* are the short time-scale data from Fig. 4.2 (shown for comparison). The data from GX 339-4 are split into those from when the source was moving towards the soft state (*small, light green diamonds*) and those from when the source was leaving the soft state (*large, dark green diamonds*); the different correlations in these data are reflective of the hysteresis curve in the GX 339-4 hardness-intensity diagram. The short and long time-scale data from XTE J1118 + 480 show remarkably similar behaviour, but the long time-scale Cygnus X-1 data all show a correlation between $\dot{m}_E$ and Γ that is consistent with the short time-scale soft-state observations. This is true even of the observations from proposal P30157, which individually appear to be of hard or intermediate state. Only the hard-state observations of GX 339-4 are shown, but in the data from just before the soft-state transition I find a $L_X$/ Γ correlation that is similar to that found in the 100 ms PCA data from this source

### 4.4.3 A Comparison of the $\dot{m}_E$–Γ Correlation on Long and Short Time-Scales

In Fig. 4.8 the different behaviours of GX 339-4 and Cygnus X-1 are further emphasised by a closer examination of the data from observing programmes P30157 (Cygnus X-1; 1997 December 11 to 1998 March 12) and P95409 (GX 339-4; 2010 January 12 to 2010 April 09), with the latter again cut off at the time the source enters

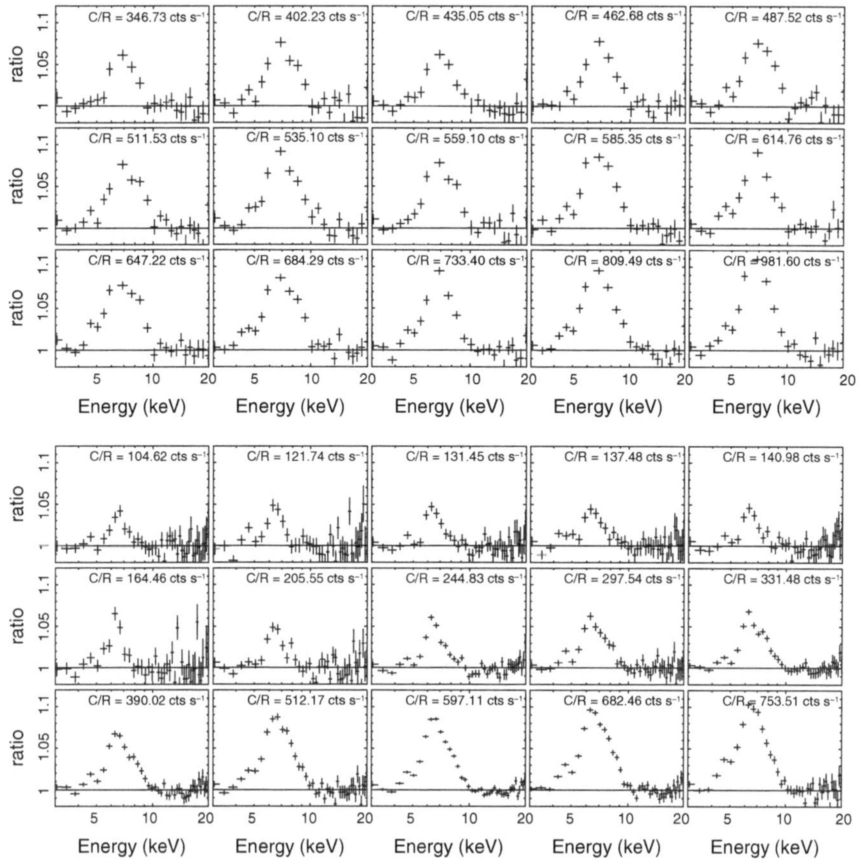


**Fig. 4.6** *Top* Hard-state X-ray spectra of GX 339-4 from the event mode data of observation 95409-01-12-01, expressed as a ratio to an absorbed power-law fit. The spectra were extracted with 1 s time-resolution, sorted by count rate and then added together in groups of 1,200. The panels are sorted by count rate, which is also displayed in the *top right-hand* corner. The iron line can clearly be seen as an excess of photons at about 6 keV, and shows a slight increase in equivalent width with count rate. *Bottom* Hard-state spectra of GX 339-4, but this time extracted from the standard-2 mode data with ~32 h time-resolution. There data are all dated from between 2010 January 12 and 2010 April 09, during which time the source was in the hard state and showed a strong $\dot{m}_E$–Γ correlation. The increase in the strength of the iron line with the count rate is much clearer in these data than in the 1 s time-resolution data, which is perhaps a consequence of the latter covering a much smaller range of accretion rates. The increase in the iron line equivalent width is unexpected, and is inconsistent with the results found for Cygnus X-1 in Chap. 3 and for GX 339-4 by Nowak et al. (2002). I gratefully thank Diego Altamirano for verifying my reduction of these data

the intermediate state. The plot shows the long time-scale standard-2 data (large purple and blue crosses) alongside the short time-scale data from a small selection of observations (open circles). In the hard state of Cygnus X-1 the change in Γ in response to a change in $L_X$ (i.e. the temperature change of the corona) is very different on 100 ms time-scales to that found on ~7 day time-scales, whereas in GX 339-4

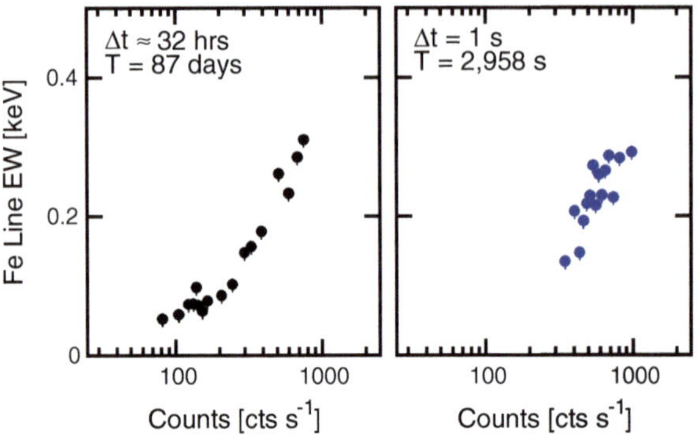


**Fig. 4.7** The equivalent width of the iron line versus count rate in GX 339-4, shown in the standard-2 data with ~32 h time-resolution (*left*) and in the event mode data with 1 s time-resolution (*right*)

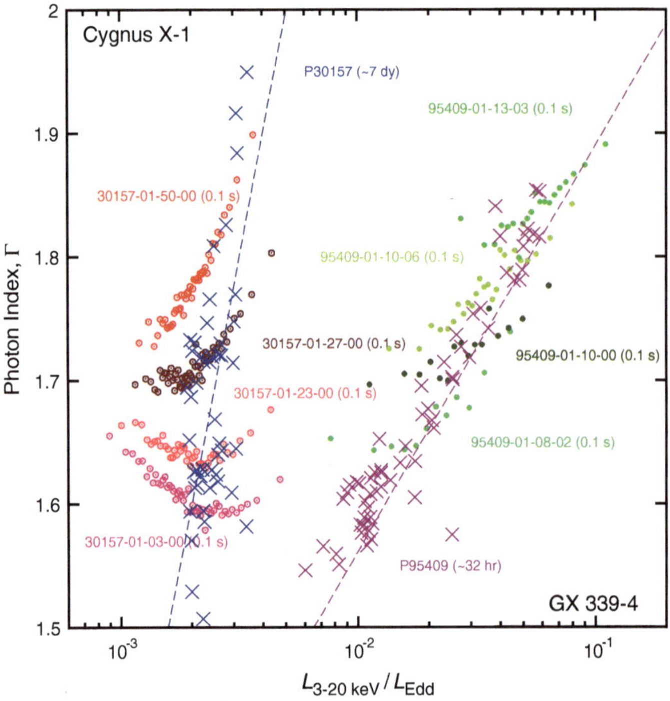


**Fig. 4.8** Long and short time-scale date from two observing programmes of Cygnus X-1 (P30157, hard and intermediate state) and GX 339-4 (P95409, hard state). The response of Γ to changes in $\dot{m}_E$ in the hard state of Cygnus X-1 is completely different on 100 ms time-scales to that found on ~7 day time-scales, while in the intermediate state there appears less difference. In GX 339-4 we find a much more similar gradient on both time-scales (though slightly less steep in the 100 ms data)

the $\dot{m}_E$–Γ distribution is very similar on both time-scales (though the gradient is slightly less steep in the 100 ms data). In Cygnus X-1 we also find that the $\dot{m}_E$–Γ correlation on 100 ms time-scales changes considerably as the source softens, with the gradient of the positive correlation increasing. In GX 339-4 the $\dot{m}_E$–Γ correlation in the 100 ms data appears exactly the same in all the observations examined.

The different $\dot{m}_E$–Γ correlations found in the long and short-time scale data from Cygnus X-1 could perhaps be attributed to physical changes in the accretion flow on long periods of time. When observed over long time-scales, the positive $\dot{m}_E$–Γ correlation (which is always due to Compton-cooling of the corona) is more likely to be affected by changes in the accretion disc geometry, such as variations in the truncation radius. A change in the truncation radius not only increases the temperature and luminosity of the inner edge of the disc (from where the seed photons for Comptonisation are generally supplied), but also increases the proportion of disc photons that intersect the Comptonising region in the centre. Therefore, an increase in flux combined with a decrease in the truncation radius could easily be responsible for creating a stronger cooling effect than that would be possible with just the flux increase on its own. Over much shorter time-scales (when there may not be sufficient time for these physical changes in the accretion disc to occur) the correlation is simply the result of changes in disc luminosity due to accretion rate variations.

It is also interesting to see in the 100 ms Cygnus X-1 observations that the transition between an anti-correlation and a positive correlation at $\dot{m}_{crit}$ moves to lower accretion rates as the source becomes softer. It can clearly be seen that there is a range of accretion rates at which Γ is positively correlated with $\dot{m}_E$ in observation 30157-01-50-00 and anti-correlated in observation 30157-01-03-00 and 30157-01-23-00, although it should be noted that changes in the bolometric correction between observations could have a significant effect on these results. As both $\dot{m}_E$ and Γ increase over long time-scales the truncation radius of the disc moves inwards and the balance of seed photon supply for Comptonisation shifts from the RIAF to the disc. However, rapid changes in the flux on 100 ms time-scales do not result in similar changes in the truncation radius and $\dot{m}_{crit}$ gradually evolves with the hardness ratio over long time-scales. This result can therefore also be interpreted in the context of a fixed (or only slowly varying) truncation radius on short time-scales.

The count rate/Γ CCFs generated from the same four Cygnus X-1 observations and three of the GX 339-4 observations are shown in Fig. 4.9 (the fourth GX 339-4 observation, 95409-01-08-02, was too faint for this analysis). The CCFs show the positive correlation in Cygnus X-1 growing in strength as the source softens, and at the same time the width of the peak becoming narrower. This result is consistent with the findings of Chap. 3 and can be attributed to a decreasing truncation radius as the source softens. The CCF of the softest GX 339-4 observation (95409-01-13-03) reveals a peak with a similar width to the softest Cygnus X-1 observations (which is consistent with the similar hardness ratios of these observations) but a weak strength and high degree of asymmetry that are more consistent with the hardest observations of Cygnus X-1.

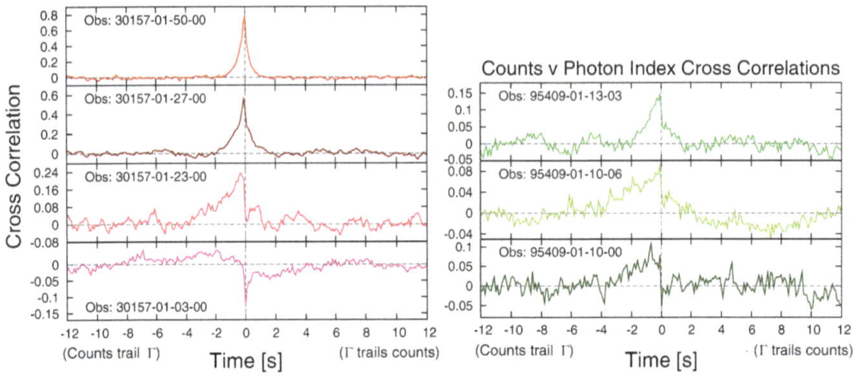


**Fig. 4.9** Count rate/Γ Cross-correlation functions for the four Cygnus X-1 observations and three of the GX 339-4 observation shown in Fig. 4.8 (GX 339-4 observation 95409-01-08-02 was too faint for this analysis). For each source the softest (and highest accretion rate) observations are shown at the *top* and the hardest (and lowest accretion rate) observations at the *bottom*. The Cygnus X-1 data show the positive correlation narrowing as the source softens, which is consistent with a decreasing truncation radius. The softest GX 339-4 observation has a peak width similar to that found in the softest Cygnus X-1 CCFs (note that the FWHM of the peak scales with the hardness ratio—see Chap. 3), but the weak strength of the peak and its high degree of asymmetry are more consistent with the hard-state observations of Cygnus X-1

### 4.4.4 Hardness-Intensity Diagrams

The different hard-state behaviour of Cygnus X-1 and GX 339-4 is clearly illustrated when hardness ratio is plotted against count rate for these sources (Fig. 4.10). GX 339-4 displays the distinctive q-shaped hysteresis curve that is always traversed

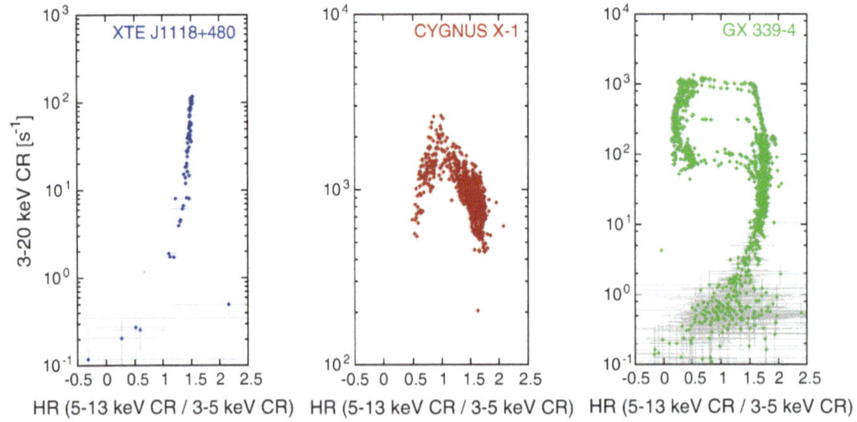


**Fig. 4.10** Hardness intensity diagrams (based upon the ratio of count rates in the 5–13 keV and 3–5 keV bands) for XTE J1118+480, Cygnus X-1 and GX 339-4, based upon all available *RXTE* standard-2 data. Whereas the XTE J1118+480 plot resembles the hard tail of the GX 339-4 plot, Cygnus X-1 enters the soft state without showing the same q-shaped hysteresis curve that is found in GX 339-4

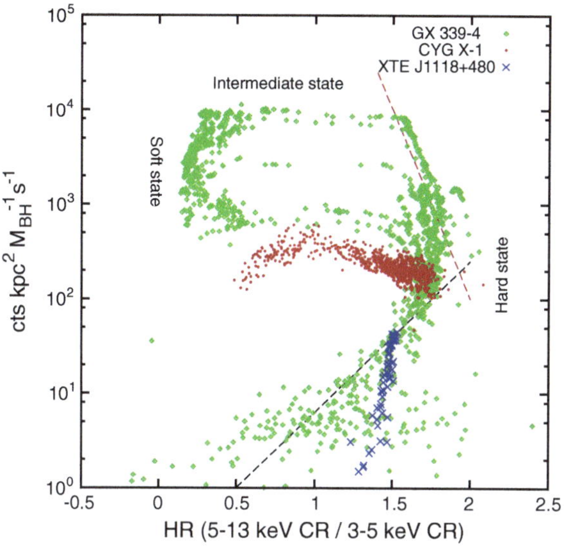


**Fig. 4.11** Hardness intensity diagrams with the count rate corrected for distance and black hole mass ($CR \times d^2/M_{\rm BH}$). A similar correction is applied to the flux in order to estimate the accretion rate, and therefore the $y$-axis on this plot should be roughly proportional to accretion rate. The labels show the hard, intermediate and soft states of GX 339-4 and the separate 'harder-when-brighter' (*black line*) and 'softer-when-brighter' (*red line*) behaviour of the hard state of this source

in an anti-clockwise direction; the hard-state (which is represented by a curve that sweeps from the bottom-left of the plot to the top-right) shows a hardening with increasing count rate at low luminosities and a softening with increasing count rate at high luminosities, which is analogous to the switch in $\dot{m}_{\rm E}$–Γ correlation. In contrast, Cygnus X-1 shows only a single path between the hard and soft states, and the extreme changes in hardness ratio in response to only small changes in count rate appear more closely related to the intermediate states of GX 339-4 than the hard states.

In order to compare the hardness-intensity diagrams of GX 339-4, Cygnus X-1 and XTE J1118+480 I choose to apply a correction factor to the count rates, such that $CR_{\rm corr} = CR \times (d^2/M_{\rm BH})$ (see Fig. 4.11). This term is proportional to the factor that would be used to convert from $F_{\rm X}$ to $L_{\rm X}/L_{\rm Edd}$, and therefore $CR_{\rm corr}$ is roughly proportional to accretion rate. The plot also indicates the hard, intermediate and soft states of GX 339-4, and highlights the separate 'harder-when-brighter' and 'softer-when-brighter' behaviour of the hard state of this source. Although this plot does not include bolometric corrections, which may be different in each source and also heavily dependent upon the spectral state, the Cygnus X-1 data appear to be most strongly associated with the path taken by GX 339-4 as it exits the soft state at around the $m_{\rm crit}$ accretion rate.

### 4.4.5 An Alternative Interpretation of the Short Time-Scale Anti-correlation in Cygnus X-1

In Figs. 4.2, 4.3 and 4.4 I showed that in many of the hard-state Cygnus X-1 observations $\Gamma$ can be anti-correlated with $\dot{m}_E$ at low accretion rates and positively correlated at high accretion rates. In Fig. 4.4 I further showed using CCFs that the source continuously crosses the boundary between these two accretion rate regimes on time-scales of less than one second. Throughout this chapter I have associated this anti-correlated component with Comptonisation in a RIAF, and this mechanism is undoubtedly responsible for producing the anti-correlation in XTE J1118+480. However, in Cygnus X-1 the very high speed with which the source appears to switch between a correlation and an anti-correlation precludes the possibility that changes in the disc truncation radius are responsible, and it therefore seems worthwhile including a discussion of the validity of this result along with a summary of my efforts to investigate alternative explanations.

One obvious solution to the problem would be to invoke the presence of a soft and constant (or only slowly varying) component. Under this scenario, a reduction in flux from the power-law component on short time-scales would increase the relative strength of the fixed component and cause the spectrum to soften. The very rapid change between a positive correlation and anti-correlation would therefore be naturally explained, as would their presence at high and low accretion rates respectively. The idea of a soft, fixed component immediately suggests an accretion disc and throughout my research I have therefore spent some considerable time attempting to fit my spectra with disc components that are able to explain the anti-correlation in Cygnus X-1.

In general I have found that a soft, fixed component can very nicely explain the Cygnus X-1 results from Fig. 4.8, but an attempt to do this using a disc black body component (DISKBB) in XSPEC requires extremely unphysical model parameters. For instance, the spectrum with the lowest flux amongst the summed spectra of the hard-state observation 30157-01-23-00 can be fitted with a power-law with $\Gamma \sim 1.6$ and a disc black body component with inner radius temperature of 1.38 keV and normalisation of 5.3. Once this model was applied to the rest of the summed spectra from this observation then the anti-correlation was indeed found to disappear. However, this disc temperature is almost three times as high as that found in the soft-state observations 10412-01-01-00, 10412-01-02-00 and 10412-01-05-00 and, even worse, implies a truncation radius well below $1\,R_G$ (which, in Cygnus X-1, is $\sim$22 km). In the soft state, the spectral fits found the disc normalisation to be between 20,000 and 26,000 (implying an inner radius of approximately 28–32 km), which (given the non-relativistic nature of the model) is reasonably consistent with a disc extending inwards to the innermost stable orbit. Consequently, when the disc component normalisation was forced to be at least 20,000 whilst fitting observation 30157-01-23-00 (which is essentially an insistence that the truncation radius is at least as large as that found in the soft state) then the spectral fits forced the temperature to be much lower ($\sim$0.25 keV) and the disc component was no longer able to explain the anti-correlation found in the data at low accretion rates.

A more viable alternative for fitting the soft component is perhaps by using an additional power-law; unlike a disc component, a power-law has parameters that provide no physical interpretation with which to constrain the fit. Using the same hard-state observation, 30157-01-23-00, it was found that the addition of a fixed power-law with $\Gamma = 2.27$ results in the disappearance of the anti-correlation, although the gradient of the $\dot{m}_E$–$\Gamma$ correlation does become considerably less steep at low accretion rates (perhaps suggesting that the parameters of this new component require some tweaking). If a fixed power-law is indeed present in the data then I can currently offer no physical interpretation of this component, but suggest that the study of rapid spectral variability could potentially be a more effective tool for further examination than spectral fitting alone.

Possibly the most convincing reason for rejecting the fixed, soft component hypothesis is provided by the CCF and ACF analysis of Chap. 3. If a soft, fixed spectral component were responsible for producing the anti-correlation then the width of this component in the CCF should be expected to be comparable to the width of the peak in the ACF (since $\Gamma$ should be varying at the same speed as the flux). Figure 4.12 shows a comparison of the ACF and $\Gamma$ versus count rate CCF of the hard-state Cygnus X-1 observation 20175-01-02-00, and reveals that the width of the peak in the ACF, although comparable to the width of the positive peak in the

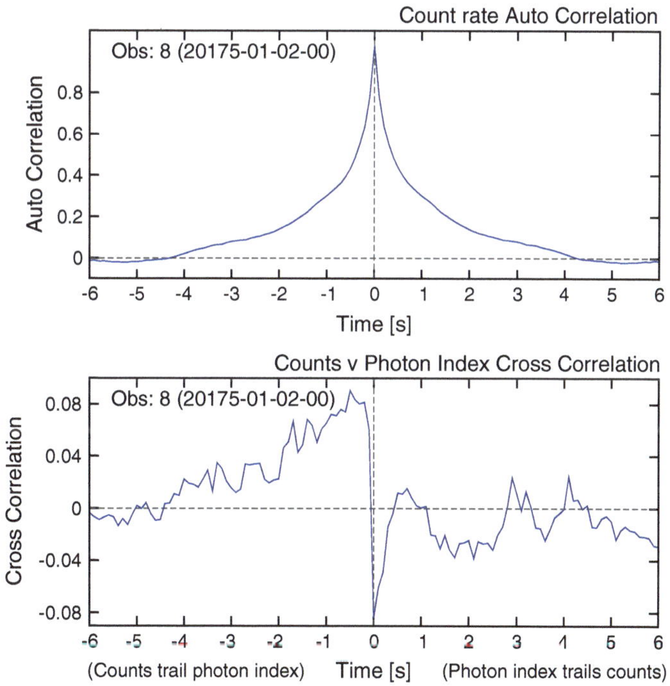


**Fig. 4.12** A comparison of the ACF (*top*) and $\Gamma$ versus count rate CCF (*bottom*) of Cygnus X-1 observation 20175-01-02-00. The width of the peak in the ACF is comparable to the positive peak in the CCF, but considerably wider than the negative peak in the CCF

CCF, is considerably greater than the width of the narrow anti-correlated peak in the CCF. It is therefore difficult to interpret this component as being a direct consequence of rapid flux changes and it was for this reason that the possibility of a fixed, soft component was originally not considered in Chap. 3.

## 4.5 The $\dot{m}_E$–Γ Relationship in AGN

The five AGN sources included in my sample have been analysed in a similar way to the XRBs, with the main difference being that, for the AGN, it is the spectra from the complete standard-2 observations that are added together and not the binned PCA spectra. The spectra are sorted by count rate, grouped together into groups of ~15–20 and then added together. In doing this I have chosen to group spectra only with others from the same proposal, and spectra with a different proposal ID are processed entirely separately. When adding standard-2 datasets it is also necessary to generate new background spectra and RMF files by using the QSUBBACK and QADDRMF parameters of the ADDSPEC tool.

In the following subsections I discuss each of my five AGN sources separately.

### *4.5.1 NGC 1052*

The LINER 1.9 galaxy NGC 1052 is located at a distance of 17.8 Mpc and hosts a black hole of mass $1.8 \times 10^8 \, M_\odot$. This source is characterised by an unusually flat X-ray spectrum, and shows evidence of possible heavy absorption by a patchy absorber with a column density close to being Compton-thick (Guainazzi and Antonelli 1999; Weaver et al. 1999). This source was also the first LINER galaxy in which polarised broad emission lines were detected (Barth et al. 1999). In Chap. 2 I listed NGC 1052 as a possible Compton-thick candidate, but this has never been firmly established despite the detection of polarised lines. The *Chandra* spectra reveal that the accretion rate is approximately $7.2 \times 10^{-3}$ % of the Eddington limit, but the true value may be at least an order of magnitude higher if the source is indeed Compton thick.

The lightcurve (Fig. 4.13, top panel) reveals the source to be extremely faint, with a count rate that places it on the very limit of the detection capability of *XTE*'s PCA instrument. The questionable reliability of the background subtraction process in very faint sources means that these data should be treated with a certain level of distrust. The measurements of Γ shown on the same plot suggest a much softer spectrum than that which was generated from *Chandra* observation 5910 (shown in Appendix A) or that which was observed by Guainazzi and Antonelli (1999) or Weaver et al. (1999), which raises further questions over the reliability of these data. When the luminosity ($L_{3-20\,\mathrm{keV}}/L_{\mathrm{Edd}}$) is plotted directly against Γ (Fig. 4.13, bottom; note that prior to fitting in XSPEC the spectra were sorted by count rate and added together in groups of 7) there appears to be no correlation between the two properties (a Spearman rank test gives $r_s = -0.208$, and therefore fails to find a correlation at the 1-$\sigma$ level of significance).

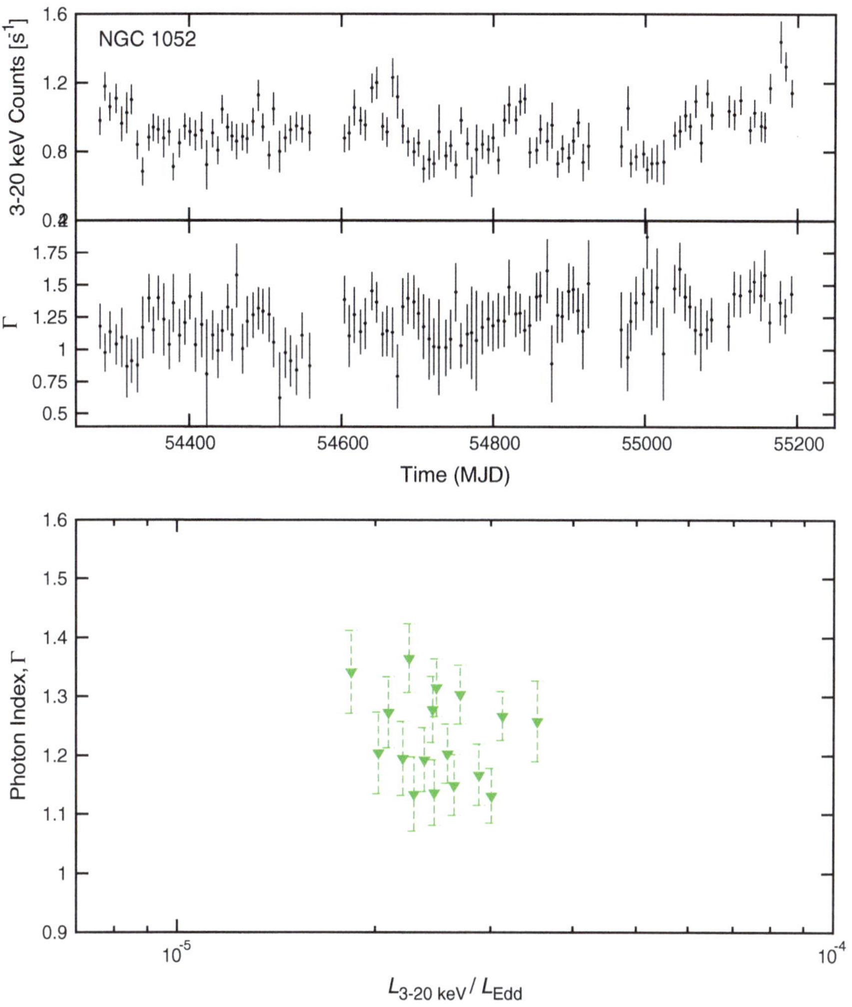


**Fig. 4.13** ***Top*** 3–20 keV count rate and photon index as a function of time for the LINER-1.9 galaxy NGC 1052. *Bottom* Accretion rate (using $L_{3-20\,\mathrm{keV}}/L_{\mathrm{Edd}}$ as a proxy) versus photon index for NGC 1052. There appears to be no correlation between these properties for this source, and it is likely that variable absorption plays an important rôle in the spectral changes observed in this source

## *4.5.2 NGC 3516*

NGC 3516 is a type-1.2 Seyfert galaxy located at a distance of 38.9 Mpc and hosts a black hole of mass $9.0 \times 10^7\,M_\odot$. The *Chandra* spectra (see Appendix B) take the form of a power-law, with a relatively complex soft excess below ~1 keV and a narrow iron line at ~6.4 keV. The average accretion rate obtained from the *Chandra* observations was found to be 7.9 % of the Eddington limit, but this value varied by

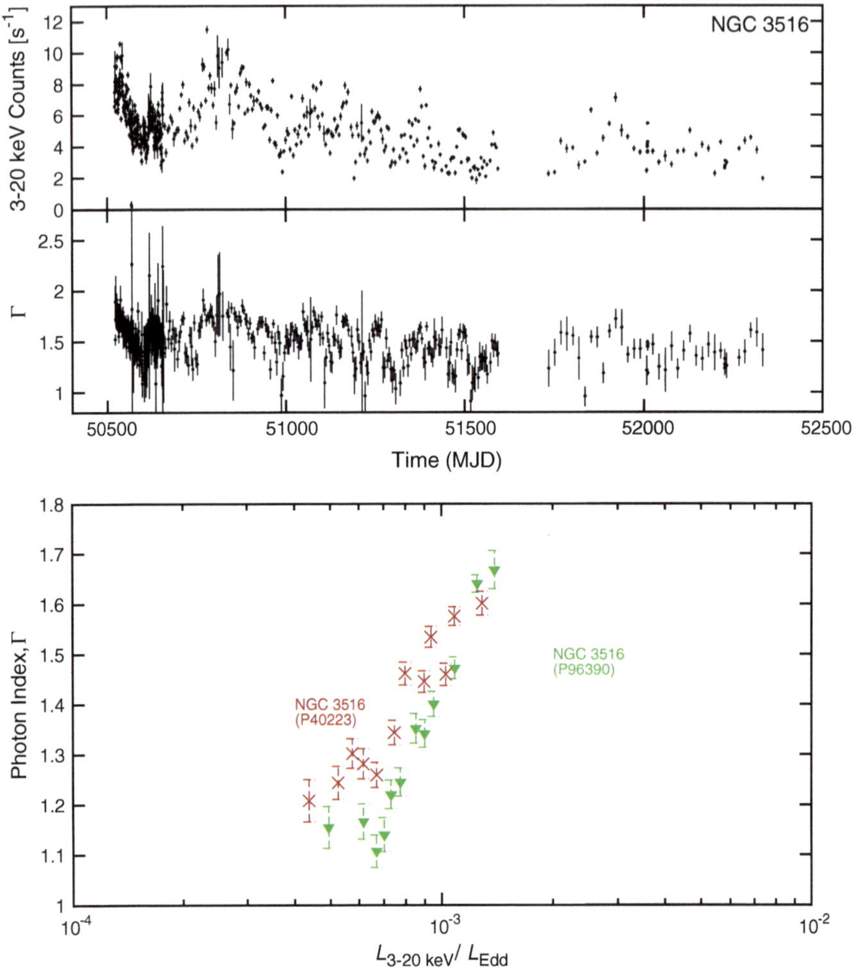


**Fig. 4.14** *Top* 3–20 keV count rate and photon index as a function of time for the Seyfert-1.2 galaxy NGC 3516. The correlation between these properties can clearly be seen in several places. *Bottom* Accretion rate (using $L_{3-20\,\rm keV}/L_{\rm Edd}$ as a proxy) versus photon index for NGC 3516, showing a strong correlation in both datasets

a factor of up to 7.5 between observations. Some of the *Chandra* observations also suffer from the effects of pile-up.

Figure 4.14 (top) shows the lightcurve (measured in counts per second in the 3–20 keV range) and photon index between 1997 February and 2002 August, which represents 498 of the 624 *XTE* observations that have been reduced for this source. A positive correlation between these properties would be expected from a source with an accretion rate above one per cent of Eddington, and such a correlation is clearly evident in this plot. The data from proposals P40223 (96 observations dating from 1999 January 1 to 2000 February 19) and P96390 (90 observations dating from

2010 December 31 to 2011 December 30) have been processed using the method described above (i.e. their spectra have been sorted by accretion rate and then added to improve the signal-to-noise ratio), and the relationship between $L_{3-20\,\mathrm{keV}}/L_{\mathrm{Edd}}$ and Γ is shown in Fig. 4.14 (bottom). A Spearman rank test finds that $L_{3-20\,\mathrm{keV}}/L_{\mathrm{Edd}}$ and Γ are correlated in both data sets at a 3-$\sigma$ level of significance ($r_s = 0.937$ in both cases).

### 4.5.3 NGC 3998

NGC 3998 is a type-1.9 LINER galaxy located at a distance of 21.6 Mpc and hosts a black hole of mass $7.3 \times 10^8\,M_{\odot}$. The *Chandra* spectra of this source are amongst the simplest in form of any I have studied in this thesis, showing only an unabsorbed power-law and lacking any thermal component or iron line. The accretion rate was found to be $4.0 \times 10^{-3}\,\%$ of the Eddington limit, which places the source firmly within the range from which we would expect advection dominated accretion to operate.

The lightcurves of both count rate and Γ (Fig. 4.15, top) are considerably more noisy than those of any other AGN studied in this chapter, and show less evidence of genuine variability. However, in several places (especially the peak in the count rate at around MJD 55 610) an anti-correlation between count rate and Γ is clearly seen in the data. This anti-correlation becomes much clearer when $L_{3-20\,\mathrm{keV}}/L_{\mathrm{Edd}}$ is plotted against Γ (Fig. 4.15, bottom), and reveals the gradient to be very steep. A Spearman rank test finds a correlation at the 3-$\sigma$ level of significance ($r_s = -0.840$). It should be noted, however, that this source is extremely faint, and therefore the background subtraction process may not be completely reliable. The lack of an iron line in my spectral fits is consistent with previous *XMM* studies of this source (e.g. Ptak et al. 2004), which have found neither an iron line nor any reflection features; NGC 3998 is therefore best described using a RIAF model, with any accretion disc truncated at some large radius.

### 4.5.4 NGC 5548

NGC 5548 is a type-1.5 Seyfert galaxy located at a distance of 67.0 Mpc and hosts a black hole of mass $6.1 \times 10^8\,M_{\odot}$. The *Chandra* spectra of this source are heavily affected by pile-up at energies greater than ~6 keV, but otherwise are consistent with a power-law that shows only a weak excess of absorption above the Galactic level. The accretion rate was estimated to be $1.6 \times 10^{-1}\,\%$ of the Eddington limit.

The lightcurve of this source (Fig. 4.16, top) covers almost the entire era of *RXTE* operations, with only two gaps in the data centred on MJD 52,900 and MJD 55,000. The source can be seen to vary in count rate by a factor of ~10 over long time-scales but, interestingly, shows very little variation in Γ that could correspond with these long time-scale changes in count rate. For instance, the source shows a broad (~400 days) peak in count rate centred on MJD 51,000, but I see no significant change in Γ during this time.

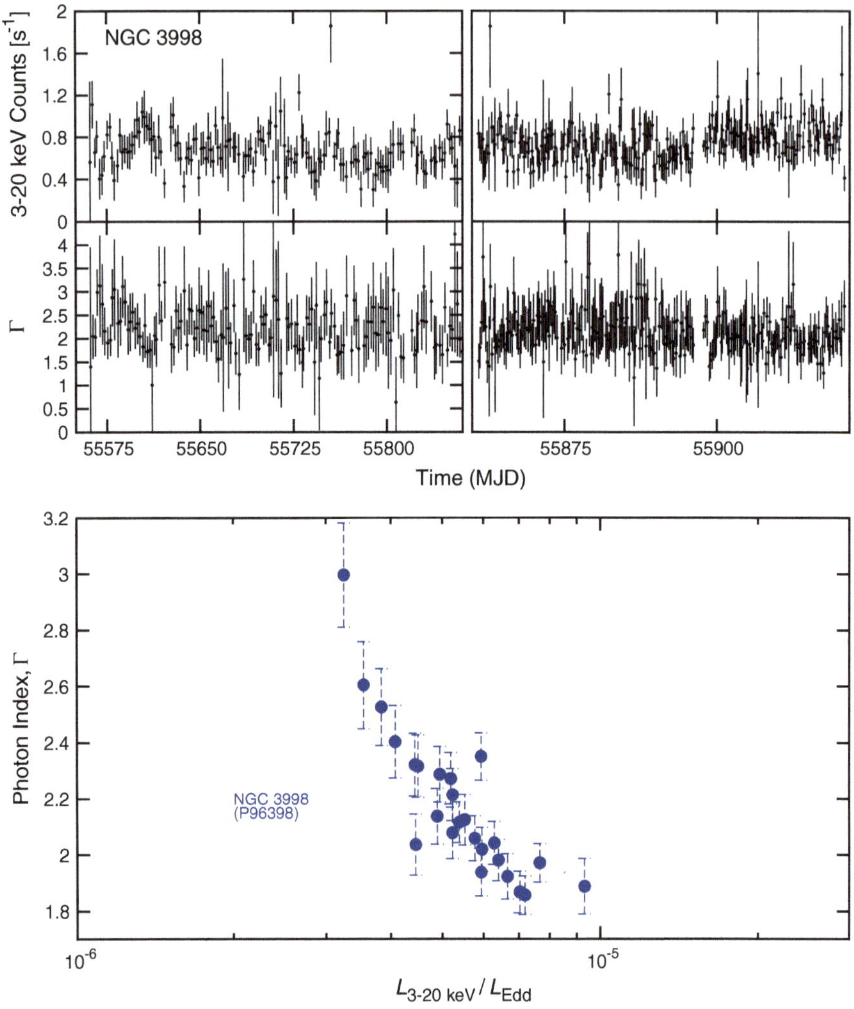


**Fig. 4.15** *Top* 3–20 keV count rate and photon index as a function of time for the LINER-1.9 galaxy NGC 3998. The separation of the lightcurves into *left* and *right* panels is to facilitate a change in scale of the $x$-axis, which was required to reflect a change in the sampling frequency of the observations. *Bottom* Accretion rate (using $L_{3-20\,\mathrm{keV}}/L_{\mathrm{Edd}}$ as a proxy) versus photon index for NGC 3998, showing a very steep anti-correlation

Despite the lack of any obvious correlation between count rate and Γ in the lightcurves, the plot of $L_{3-20\,\mathrm{keV}}/L_{\mathrm{Edd}}$ against Γ for this source (Fig. 4.16, bottom) suggests that the two properties are indeed positively correlated (a Spearman rank test finds that $r_s = 0.855$, and therefore a correlation at the 3-$\sigma$ level). This plot shows the data from 335 *XTE* observations (all from proposal P50155, and dating from between March 2000 and August 2001), with the spectra grouped by count rate into 23 groups of ∼15. The strength of the correlation is slightly weaker than that

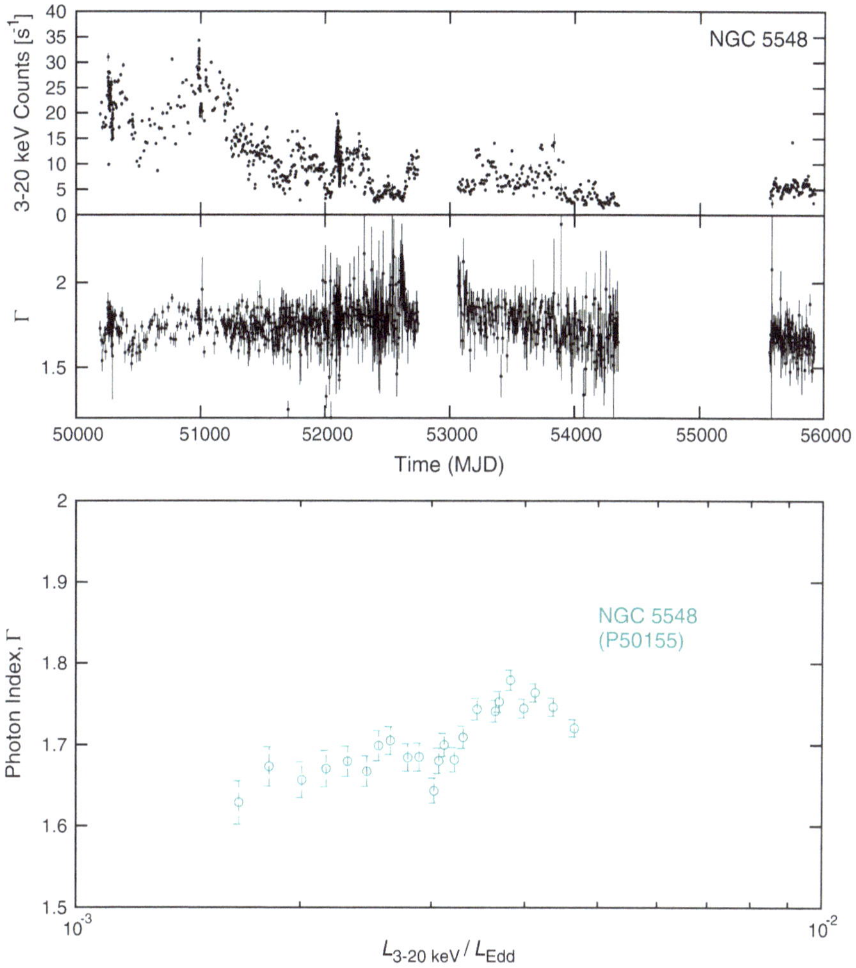


**Fig. 4.16** *Top* 3–20 keV count rate and photon index as a function of time for the Seyfert-1.5 galaxy NGC 5548. *Bottom* Accretion rate (using $L_{3-20\,\mathrm{keV}}/L_{\mathrm{Edd}}$ as a proxy) versus photon index for NGC 5548

found in the other Seyfert galaxy in my sample, NGC 3516, and the gradient is also less steep.

### 4.5.5 NGC 7213

NGC 7213 hosts a Seyfert/LINER hybrid nucleus powered by a $9.8 \times 10^7\, M_\odot$ black hole, and is located at a distance of 22.1 Mpc. This source has already been shown to demonstrate an anti-correlation between flux and Γ (Emmanoulopoulos et al. 2012), and has been included in my sample in an attempt to reproduce this result and therefore validate my methods.

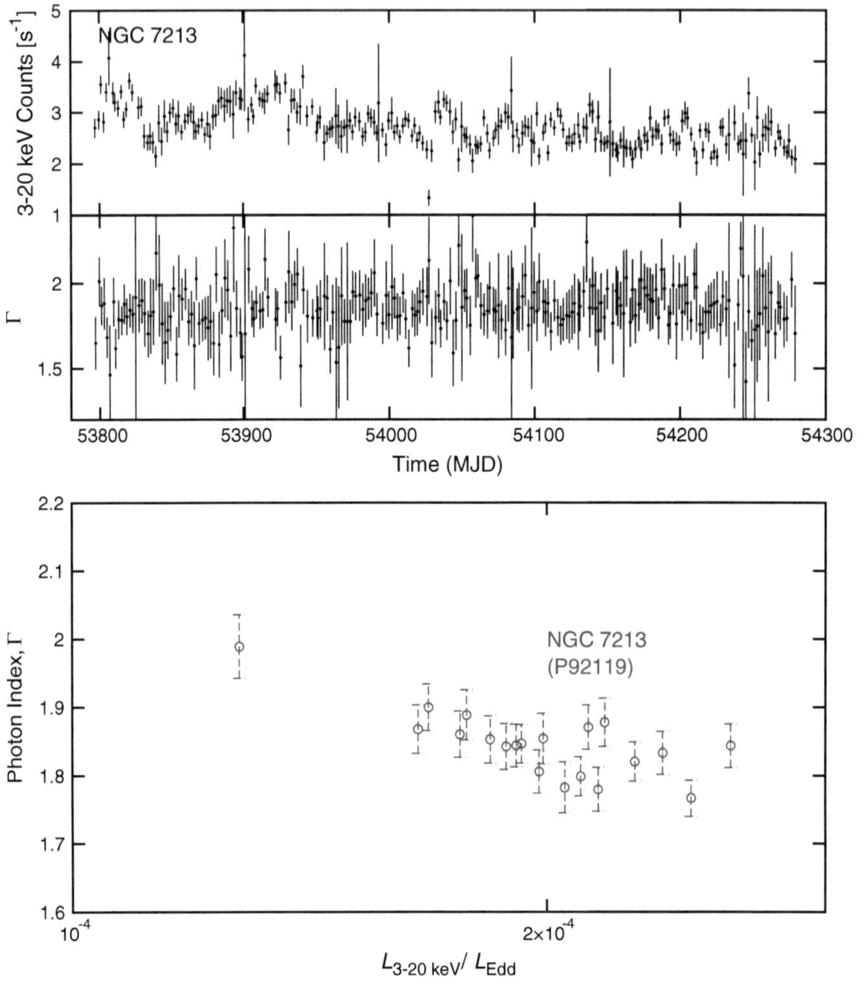


**Fig. 4.17** *Top* 3–20 keV count rate and photon index as a function of time for the LINER/Seyfert-1.5 galaxy NGC 7213. *Bottom* Accretion rate (using $L_{3-20\,\mathrm{keV}}/L_{\mathrm{Edd}}$ as a proxy) versus photon index for NGC 7213. The data are from spectral fits to 238 *XTE* observations (all from proposal P92119, and dating from between March 2006 and June 2007), grouped by count rate into 20 groups of ~12. The plot reveals a weak anti-correlation between the properties

The lightcurve of NGC 7213 (Fig. 4.17, top) shows the source to vary by a factor of less than two over a period of ~16 months, while the measurements of $\Gamma$ also show little variability over the same period. Plotting $L_{3-20\,\mathrm{keV}}/L_{\mathrm{Edd}}$ directly against $\Gamma$ (Fig. 4.17, bottom) reveals a weak anti-correlation (consistent with the findings of Emmanoulopoulos et al. 2012). A Spearman rank test finds a correlation at the 2-$\sigma$ level of significance ($r_s = -0.618$), which persists even if the outlying data point at $\Gamma \approx 1.99$ is excluded (in which case $r_s = -0.554$).

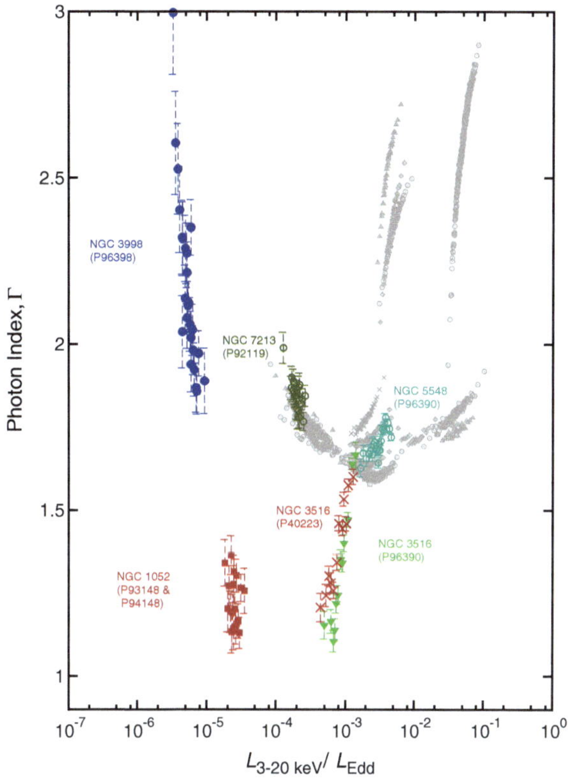


**Fig. 4.18** Plot of $L_X$ versus Γ for the five AGN sources, with the XRB sources (*grey points*) shown for comparison. The data from NGC 5548 and NGC 7213 appear to be fairly consistent with that from the hard- and intermediate-state XRB observations, whereas NGC 3516 and NGC 3998 show correlations with much steeper gradients. The observations of NGC 1052 appear to show no correlation between $L_X$ and Γ

### *4.5.6 Summary of AGN Results*

In Fig. 4.18 I show the results for all five of my AGN sources on the same plot, along with the results of the 100 ms study of my XRB sources (grey points) for comparison. In general, we see a positive correlation between $L_X$ and Γ in the two brightest sources (the Seyfert galaxies NGC 3516 and NGC 5548), and an anti-correlation in two of the fainter sources (NGC 3998 and NGC 7213); the data from NGC 1052 show no correlation.

Whereas the presence of a positive correlation in the bright sources and an anti-correlation in the faint sources is not a surprise, the AGN data are still difficult to understand. Unlike the XRBs, for which some observations show a non-linear relationship between $L_X$ and Γ, the AGN data can generally be fitted with straight lines, and the best fits to NGC 3516 and NGC 3998 have much steeper gradients than any of the hard-state XRB observations, indicating quite strong changes in spectral index in response to only slight changes in source flux. Furthermore, three of the AGN are found to display quite extreme values of $\Gamma$ that can be as low as 1.2 (NGC 1052 and NGC 3516) and as high as 3 (NGC 3998). Although the very hard spectra of NGC 1052 can possibly be attributed to heavy absorption, there is little evidence in the literature to support a similar conclusion for NGC 3516.

Of the five AGN sources in the sample, only NGC 5548 and NGC 7213 show behaviour that is consistent with that found in the hard- and intermediate-state XRBs. NGC 7213, for example, has a spectral index, an accretion rate and a $L_X/\Gamma$ correlation that are all very similar to the observations of XTE J1118+480, and NGC 5548 is very similar to the intermediate-state observations of Cygnus X-1 (especially observation 30157-01-27-00). The very steep correlations found in NGC 3516 and NGC 3998 are more consistent with the soft-state XRB observations, but the low value of $\Gamma$ in NGC 3516 and the anti-correlation in NGC 3998 would certainly appear to preclude this possibility.

## 4.6 Summary of the Main Observational Results

- By extracting spectra from thousands of 100 ms sections of *RXTE* data, and adding together those with similar count rates, I show here for the first time the presence of an anti-correlation between $\dot{m}_E$ and $\Gamma$ in a single observation of a X-ray binary system (XTE J1118+480). This anti-correlation has exactly the same gradient as that found for the same source on long time-scales (weeks and months).
- A similar analysis of the hard state of GX 339-4 (only accretion rates above $\sim$1 % were examined) reveals a positive $\dot{m}_E/\Gamma$ correlation that also has a similar gradient on two vastly different time-scales (seconds and months). The data also reveal an increase in the equivalent width of the iron line with accretion rate on both time-scales, which is inconsistent with the results from Cygnus X-1 in Chap. 3.
- In contrast to the other two XRB sources, the analysis of Cygnus X-1 reveals a very different behaviour on short time-scales to that found on much longer time-scales. On long time-scales, Cygnus X-1 shows a very steep $\dot{m}_E/\Gamma$ correlation that is similar to that found in GX 339-4 as it exits the soft state and moves towards the hard state. On short time-scales, and especially in the hard state, the gradient of the $\dot{m}_E/\Gamma$ correlation is much less steep, and is more similar to that found in the hard state of GX 339-4.
- Many of the hard-state Cygnus X-1 observations reveal the presence of both an anti-correlation and a positive correlation within the same observation. This much was already suspected from the CCFs of Chap. 3, but these data confirm that the anti-correlation dominates when the count rate is low and the positive correlation when the count rate is high. The lightcurves from these observations suggest that the source continuously switches (on sub-second time-scales) between an anti-correlation and a positive correlation, and CCFs between count rate and $\Gamma$ further suggest that the width of the positive correlation (perhaps associated with the truncation radius of the accretion disc) does not change significantly during these times.

- Many of the observations from Cygnus X-1 observing programme P30157 show both an anti-correlation and a positive correlation, and an examination of these data reveal that the critical accretion rate ($\dot{m}_{\rm crit}$) at which the source switches from an anti-correlation to a positive correlation appears to move to lower accretion rates as the source softens.
- Of the five AGN sources examined, four show a correlation between $\dot{m}_{\rm E}$ and $\Gamma$. Whereas two of these sources (NGC 5548 and NGC 7213) have $\Gamma$ distribution that are similar to those found in the XRBs, two further sources (NGC 3516 and NGC 3998) show much steeper correlations that have large gradients which are similar in magnitude to the long-term Cygnus X-1 data or the short time-scale soft-state Cygnus X-1 data.

## 4.7 Discussion

In this chapter I have introduced a technique for directly comparing the spectral variability of XRBs with that of AGN, taking into account the effect of black hole mass on the variability timescale. By adding together spectra with similar count rates I can hopefully reveal any intrinsic correlation between accretion rate and $\Gamma$, whilst allowing any other random variation (which may affect the measurement of $\Gamma$ in any individual spectrum) to cancel out. This technique is particularly suitable for studying the low/hard states of XRBs, in which the spectra are of a simple power-law form and lack features such as a disc component, iron line or soft excess that could be changing independently of the power-law. The addition of hard-state XRB spectra has the further advantage of increasing the low count statistics from these observations, and therefore improving both the signal-to-noise ratio and the quality of the spectral fits.

In Fig. 4.2 I showed the presence of an anti-correlation between $L_X/\Gamma$ in two observations of XTE J1118+480 (of length 12.4 ks and 13.9 ks), sampled with a time-resolution of 100 ms. Although the anti-correlation in low accretion rate XRBs has been shown before over long time-scales (e.g. Wu and Gu 2008), I believe this plot is the first time it has been shown to exist within a single observation.

More interesting, perhaps, are the observations of Cygnus X-1 from the same plot (and analysed further in Fig. 4.4) which reveal a transition from an anti-correlation to a positive correlation within a single *XTE* observation. The lightcurve in Fig. 4.4 shows that the source switches continuously from a low count rate ($<$4,000 cts s$^{-1}$), in which the CCF of count rate versus $\Gamma$ reveals an anti-correlation, to a high count rate, in which the CCF reveals a positive correlation, on time-scales of less than one second. The positive correlation peak in these CCFs is considerably broader than the anti-correlation, which supports the idea that the latter originates in the innermost regions of the accretion flow. The CCFs also indicate that the broad wings of the positive correlation peak are still present in the CCF generated from only the low count rate data, implying that this component is always present throughout the duration of this observation. Furthermore, the width of the positive correlation peak (measured at the base) remains the same in both the high and low count rate CCF. In Chap. 3 I argued that changes in the width of the positive correlation peak were caused by changes in the truncation radius of the accretion disc, and therefore the identical widths of the peaks in the high and low count rate CCFs suggest that the truncation radius of the disc is not varying significantly on very short timescales and cannot be responsible for the transition between positive correlation and anti-correlation in this observation. However, although it is entirely plausible that accretion rate perturbations can propagate inwards through the accretion flow and alter the balance of seed photon supply between the disc and the RIAF, it is not entirely clear how disc- and RIAF-dominated Comptonisation can be associated with high and low accretion rates respectively without invoking an accretion-rate-dependent truncation radius.

The way in which $\Gamma$ reacts to flux changes in hard-state XRBs shows some surprising consistencies on two very different time-scales. In Fig. 4.5 I show that the $L_{\rm X}/\Gamma$ anti-correlation found in the 100 ms PCA data from XTE J1118+480 is almost identical to the anti-correlation found in the standard-2 data spanning a period of months. Similarly, the positive correlation found in the 100 ms PCA data from GX 339-4 (observations 40031-03-02-03 and 70109-01-04-00) resembles that found in the standard-2 data spanning the period 2010 January 12 to 2010 April 09. There is, however, some suggestion of a softening in $\Gamma$ at the low accretion rate end of the GX 339-4 100 ms data (when compared to the long time-scale data), which becomes a clear anti-correlation in observation 20056-01-01-00. A similar effect is also observed in Cygnus X-1 (see also Fig. 4.3), in which an anti-correlation is observed in the 100 ms time-resolution data, is found to become much weaker in the 1.5 s time-resolution data and is not present at the same accretion rate on longer time-scales. This association of the anti-correlation with the fastest variability implies an origin that is deep within the innermost regions of the accretion flow; in addition, its tendency to be found only at low accretion rates suggests that the RIAF is responsible for producing this component.

Although the gradient of the $L_{\rm X}/\Gamma$ anti-correlation within the 100 ms data is reasonably consistent between sources, the same cannot be said of the positive correlation on long time-scales. I have already described how, on short time-scales, Cygnus X-1 can show both a negative and positive correlation within data from the same hard-state observation, and the gradients of these correlations are consistent with those found in the short time-scale data of XTE J1118+480 and GX 339-4. On long time-scales, however, Cygnus X-1 shows a much steeper, positive correlation that is similar in gradient to both the 100 ms data from the soft-state observations and the long time-scale GX 339-4 data from when the source was returning to the hard state. The reason for this difference between the two sources is connected to the accretion rate at which the source makes the transition from the hard state to the hard intermediate state ($\dot{m}_{\rm H \rightarrow HI}$), which is an order of magnitude lower in Cygnus X-1 than in GX 339-4. Unlike GX 339-4, Cygnus X-1 does not display a hysteresis effect in its hardness-intensity diagram (see Fig. 4.10) and state transitions to and from the soft state occur at the same accretion rate. In Cygnus X-1, $\dot{m}_{\rm H \rightarrow HI}$ is very close to $\dot{m}_{\rm crit}$ which means there is only a very small range of accretion rates at which the source can be in the hard state and exhibiting a positive $L_{\rm X}/\Gamma$ correlation. The fact that a positive correlation with a shallow gradient *is* observed in the 100 ms hard-state data may be further evidence in support of a fixed disc truncation radius on these time-scales (as discussed above), and, conversely, the very different behaviour observed on long time-scales may be a consequence of the accretion flow having sufficient time to restructure itself in response to changes in accretion rate (such as changes in the truncation radius).

One of the main objectives of this chapter was to investigate the $L_X/\Gamma$ relationship in such as way as to permit a direct comparison between XRBs and AGN. Unfortunately, the large number of observations of each source that are required for this analysis meant that I was restricted to studying only five targets. Although four of the five sources show correlations between $L_X$ and $\Gamma$ that are positive at high accretion rates (NGC 3516 and NGC 5548) and negative at low accretion rate (NGC 3998 and NGC 7213), only in NGC 5548 and NGC 7213 did the spectral variability behaviour appear similar to that observed in the XRBs. Of the other sources, NGC 1052 shows no clear correlation and NGC 3516 and NGC 3998 have $d\Gamma/dL_X$ gradients that are extremely steep. Although the soft-state XRB observations show similarly steep $d\Gamma/dL_X$ gradients, the hard spectrum of NGC 3516 and the low accretion rate of NGC 3998 preclude the possibility that these sources are in the soft state.

**Acknowledgments** I gratefully thank Diego Altamirano for verifying my reduction of the *RXTE* PCA standard-2 data from GX 339-4 (Fig. 4.6).

# References

Barth, A. J., Filippenko, A. V., & Moran, E. C. (1999). Polarized broad H$\alpha$ emission from the LINER nucleus of NGC 1052. *Astrophysical Journal Supplement*, *515*, L61–L64.

Constantin, A., Green, P., Aldcroft, T., Kim, D.-W., Haggard, D., Barkhouse, W., et al. (2009). Probing the balance of AGN and star-forming activity in the local Universe with ChaMP. *Astrophysical Journal*, *705*, 1336–1355.

Emmanoulopoulos, D., Papadakis, I. E., McHardy, I. M., Arévalo, P., Calvelo, D. E., & Uttley, P. (2012). The 'harder when brighter' X-ray behaviour of the low-luminosity active galactic nucleus NGC 7213. *Monthly Notices of the Royal Astronomical Society*, *424*, 1327–1334.

Gelino, D. M., Balman, Ş., Kızıloğlu, Ü., Yılmaz, A., Kalemci, E., & Tomsick, J. A. (2006). The inclination angle and mass of the black hole in XTE J1118+480. *Astrophysical Journal*, *642*, 438–442.

Gierliński, M., & Newton, J. (2006). X-ray spectral transitions of black holes from RXTE All-sky monitor. *Monthly Notices of the Royal Astronomical Society*, *370*, 837–844.

Gu, M., & Cao, X. (2009). The anticorrelation between the hard X-ray photon index and the Eddington ratio in low-luminosity active galactic nuclei. *Monthly Notices of the Royal Astronomical Society*, *399*, 349–356.

Guainazzi, M., & Antonelli, L. A. (1999). The flat X-ray spectrum of the LINER NGC 1052. *Monthly Notices of the Royal Astronomical Society*, *304*, L15–L19.

Ho, L. C., Filippenko, A. V., & Sargent, W. L. W. (1997). A search for "dwarf" seyfert nuclei. III. Spectroscopic parameters and properties of the host galaxies. *Astrophysical Journal Supplement*, *112*, 315.

Ho, L. C., Greene, J. E., Filippenko, A. V., & Sargent, W. L. W. (2009). A search for "dwarf" seyfert nuclei. VII. A catalog of central stellar velocity dispersions of nearby galaxies. *Astrophysical Journal Supplement, 183*, 1–16.

Hynes, R. I., Steeghs, D., Casares, J., Charles, P. A., & O'Brien, K. (2003). Dynamical evidence for a black hole in GX 339–4. *Astrophysical Journal Letters, 583*, L95–L98.

Maccarone, T. J. (2003). Do X-ray binary spectral state transition luminosities vary? *Astronomy & Astrophysics, 409*, 697–706.

Magdziarz, P., & Zdziarski, A. A. (1995). Angle-dependent Compton reflection of X-rays and gamma-rays. *Monthly Notices of the Royal Astronomical Society, 273*, 837–848.

McClintock, J. E., Garcia, M. R., Caldwell, N., Falco, E. E., Garnavich, P. M., & Zhao, P. (2001). A black hole greater than 6 $M_{solar}$ in the X-ray Nova XTE J1118+480. *Astrophysical Journal Letters, 551*, L147–L150.

McHardy, I. M., Koerding, E., Knigge, C., Uttley, P., & Fender, R. P. (2006). Active galactic nuclei as scaled-up galactic black holes. *Nature, 444*, 730–732.

Nowak, M. A., Wilms, J., & Dove, J. B. (2002). Coronal-temporal correlations in GX 339–4: Hysteresis, possible reflection changes and implications for advection-dominated accretion flows. *Monthly Notices of the Royal Astronomical Society, 332*, 856–878.

Orosz, J. A., McClintock, J. E., Aufdenberg, J. P., Remillard, R. A., Reid, M. J., Narayan, R., et al. (2011). The mass of the black hole in Cygnus X-1. *Astrophysical Journal, 742*, 84.

Ponti, G., Miniutti, G., Cappi, M., Maraschi, L., Fabian, A. C., & Iwasawa, K. (2006). XMM-Newton study of the complex and variable spectrum of NGC 4051. *Monthly Notices of the Royal Astronomical Society, 368*, 903–916.

Ptak, A., Terashima, Y., Ho, L. C., & Quataert, E. (2004). Testing radiatively inefficient accretion flow theory: An XMM-Newton observation of NGC 3998. *Astrophysical Journal, 606*, 173–184.

Reid, M. J., McClintock, J. E., Narayan, R., Gou, L., Remillard, R. A., & Orosz, J. A. (2011). The trigonometric parallax of Cygnus X-1. *Astrophysical Journal, 742*, 83.

Shemmer, O., Brandt, W. N., Netzer, H., Maiolino, R., & Kaspi, S. (2006). The hard X-ray spectral slope as an accretion rate indicator in radio-quiet active galactic nuclei. *Astrophysical Journal Letters, 646*, L29–L32.

Sobolewska, M. A., Papadakis, I. E., Done, C., & Malzac, J. (2011). Evidence for a change in the X-ray radiation mechanism in the hard state of galactic black holes. *Monthly Notices of the Royal Astronomical Society, 417*, 280–288.

Tremaine, S., Gebhardt, K., Bender, R., Bower, G., Dressler, A., Faber, S. M., et al. (2002). The slope of the black hole mass versus velocity dispersion correlation. *Astrophysical Journal, 574*, 740–753.

Wagner, R. M., Foltz, C. B., Shahbaz, T., Casares, J., Charles, P. A., Starrfield, S. G., et al. (2001). The halo black hole X-ray transient XTE J1118+480. *Astrophysical Journal, 556*, 42–46.

Weaver, K. A., Wilson, A. S., Henkel, C., & Braatz, J. A. (1999). X-ray emission from the prototypical LINER Galaxy NGC 1052. *Astrophysical Journal, 520*, 130–136.

Woo, J.-H., & Urry, C. M. (2002). Active galactic nucleus black hole masses and bolometric luminosities. *Astrophysical Journal, 579*, 530–544.

Wu, Q., & Gu, M. (2008). The X-ray spectral evolution in X-Ray binaries and its application to constrain the black hole mass of ultraluminous X-ray sources. *Astrophysical Journal, 682*, 212–217.

Wu, Y. X., Belloni, T. M., & Stella, L. (2010). Fast spectral variability from Cygnus X-1. *Monthly Notices of the Royal Astronomical Society, 408*, 2413–2425.

Zdziarski, A. A., Gierliński, M., Mikołajewska, J., Wardziński, G., Smith, D. M., Harmon, B. A., et al. (2004). GX 339–4: The distance, state transitions, hysteresis and spectral correlations. *Monthly Notices of the Royal Astronomical Society, 351*, 791–807.

Zdziarski, A. A., Poutanen, J., Paciesas, W. S., & Wen, L. (2002). Understanding the long-term spectral variability of Cygnus X-1 with burst and transient source experiment and all-sky monitor observations. *Astrophysical Journal, 578*, 357–373.

# Chapter 5
# Conclusions

*The main purpose of science is simplicity and as we understand more things, everything is becoming simpler.*

Edward Teller (1908–2003)

*Science may be described as the art of systematic oversimplification.*

Karl R. Popper (1902–1994)

## 5.1 Summary of Findings

In this thesis I have presented the results of an extensive *RXTE* archival study of short time-scale X-ray spectral variability in XRBs, with particular emphasis on the Comptonised component of the emission. I further attempted to examine the variability of the same Comptonised component in AGN spectra over comparable mass-scaled time-scales. In a related study I have also examined the nuclear emission from a sample of nearby galaxies through a large *Chandra* archival snapshot survey, and discussed the implications of this work towards developing an improved understanding of the nature of low luminosity AGN. In this section I shall summarise the main conclusions of this work, and suggest areas in which future research could augment these results.

### *5.1.1 The Nature of Compact Emission Regions in Nearby Galactic Nuclei*

In order to compare the spectral variability of AGN with XRBs, it is first necessary to ensure that the emission we see in nearby galactic nuclei is indeed the result of accretion onto a SMBH and not contamination from nearby sources. The safest



C. Skipper, *Fast Spectral Variability in the X-ray Emission of Accreting Black Holes*, Springer Theses, DOI 10.1007/978-3-319-09587-5_5

way to ensure a sample consists only of AGN is to select sources that have very high X-ray luminosities and broad Balmer lines in their optical spectra, but to do so would restrict the study to accretion rates well above $\dot{m}_{\rm crit}$ and prevent any opportunity of examining the expected anti-correlation at low accretion rates.

In Chap. 2 I found there to be no evidence of Compton thickness amongst sources where $L_{\rm [O\,III]} \lesssim 10^{39}$ erg s$^{-1}$, and the unified model (in which the orientation of the torus with respect to the observer determines the level of absorption) does not appear to apply to low accretion rate sources. Whereas sources with bright X-ray and [O III] luminosities show a clear distinction between Compton-thick and Compton-thin candidates when plotted on a graph of $L_{2-10\,\rm keV}$ versus $L_{\rm [O\,III]}$ (see Fig. 2.10), faint sources show neither any spectral evidence of Compton thickness nor any bimodality in their $L_{2-10\,\rm keV}/L_{\rm [O\,III]}$ ratios. The lack of Compton-thick candidates at low luminosities, combined with low column densities found in these sources, suggests the molecular torus is either absent or has weakened to a level that makes it indistinguishable from absorption on larger scales (see also Panessa and Bassani 2002).

If the torus is absent then the scatter in the $L_{2-10\,\rm keV}/L_{\rm [O\,III]}$ ratio of low luminosity AGN is unlikely to be a reliable indicator of heavy obscuration and, instead, the ionisation of the narrow line region around these sources may often be dominated by stellar processes. Further evidence of the non-AGN nature of highly ionised low-luminosity sources comes from a comparison of their $N_{\rm H}/E(B-V)$ ratios with the empirical relationship between column density and colour excess of Bohlin et al. (1978). Compton-thin Seyferts and LINERs with high X-ray to [O III] ratios ($L_{2-10\,\rm keV} > 5.53\,L_{\rm [O\,III]}$) show much larger X-ray column densities than would be expected from their optical extinction (see Fig. 2.11), presumably due to the origin of the X-ray being deep within the AGN where the photons are subjected to much greater obscuration than that which affects the narrow line region emission further out. Compton-thin Seyferts and LINERs with low X-ray to [O III] ratios ($L_{2-10\,\rm keV} \leq 5.53\,L_{\rm [O\,III]}$) show no such discrepancy between $N_{\rm H}$ and $E(B-V)$.

### *5.1.2 X-ray Spectral Variability in XRBs*

The main focus of this thesis has been upon the rapid variability of the Comptonised spectral component of XRBs, and specifically the switch in the sign of the $\dot{m}_{\rm E}$–$\Gamma$ correlation at $\dot{m}_{\rm crit}$. In Chap. 3 I showed that $\Gamma$ and count rate (acting as a proxy for $\dot{m}_{\rm E}$) are correlated on time-scales as short as 16 ms in Cygnus X-1, and the cross-correlation functions (CCFs) of these properties are highly asymmetric in the hard state. The shape of the CCFs is also highly dependent upon the hardness ratio, with a positive correlation in the softest states (attributed to Compton-cooling of the corona by disc black body seed photons), an anti-correlation in the hardest states (attributed to Comptonisation of cyclo-synchrotron seed photons in a hot, optically thin, geometrically thick RIAF), and in the remaining CCFs an almost monotonic transition with varying hardness ratio between the shape of the soft- and hard-state

CCFs. The positive correlation was found to be extremely narrow in the softest states ($FWHM \approx 140$ ms), and broadens with increasing hardness ratio until the first hint of an anti-correlation is detected (at which point $FWHM \approx 1.4$ s). The broadening of the positive correlation implies that the size of the emission region is increasing, which is consistent with the expected increase in the disc truncation radius with hardness ratio and, consequently, the increase in the size of the corona that has been hypothesised to exist above the inner region of the disc. The anti-correlation is narrow ($FWHM \lesssim 0.5$ s) and, although its strength increases with increasing hardness ratio, it shows little evidence of broadening; in the CCFs where both the positive correlation and anti-correlation are visible then the latter is at least four times narrower than the former, which suggests these components may originate in the outer and inner regions of the accretion flow respectively. The data are therefore consistent with a two-zone accretion model with an inner RIAF and an outer thin disc.

The high degree of asymmetry found in the hard-state CCFs may be connected to the presence of hard lags. If an increase in soft flux is detected slightly before an increase in hard flux then measurements of $\Gamma$ during this time would reveal a slight softening followed by a slight hardening, thereby introducing the weak asymmetric component into the CCFs. Such spectral changes can be explained if there is an extended emission region with a radial gradient in energy or temperature of the Comptonising electrons such that the higher energy electrons are closer to the black hole. An inwardly moving perturbation in the accretion rate would therefore boost the low-energy photon emission rate before that of the higher energy photons.

The switch in the sign of the $\dot{m}_E$–$\Gamma$ correlation at $\dot{m}_{crit}$ is widely believed to represent a transition between Comptonisation in a hot, optically thin, geometrically thick accretion flow at low $\dot{m}_E$ to Comptonisation of black body seed photons from a thin, optically thick accretion disc at high $\dot{m}_E$. In Chap. 4 (Fig. 4.4) I showed that this transition is happening on time-scales comparable to the sampling time-scale of 100 ms and, therefore, major changes in the geometry of the accretion flow (such as changes in the truncation radius) seem unlikely to be responsible. Instead, the observation shown in Fig. 4.4 can be better explained using a two-zone accretion model of a truncated disc (with truncation radius roughly fixed on these time-scales) with an inner RIAF, and inwardly travelling perturbations in the accretion rate are therefore able to alter the balance of Comptonised emission from the two accretion zones on time-scales as short as the viscous time-scale.

Finally, I show in Fig. 4.5 that the long-term hard-state spectral changes in GX 339-4 and XTE J1118+480 (sampled with ~1–4 dy time-resolution over a period of ~4–12 months) are remarkably consistent with those observed on 100 ms time-scales for these sources, but that this is not the case for Cygnus X-1. In the discussion at the end of Chap. 4 I suggest that the positive correlation between $L_X$ and $\Gamma$ in the 100 ms Cygnus X-1 data (which has a much steeper gradient in the long time-scale data) may be the result of a fixed disc truncation radius on 100 ms time-scales.

### 5.1.3 A Comparison with AGN

The nature of AGN-like emission from nearby galaxies can be both varied and complex, and therefore requires a little caution be exercised before any comparison is made with XRBs. AGN are usually found in much denser environments than their stellar-mass equivalents, and their X-ray spectra often show both complex absorption and the presence of additional components (such as a soft excess or emission spectrum from a hot, diffuse gas). In Chap. 4 I examined the long-term spectral variability of five AGN and found that only two (NGC 5548 and NGC 7213, both type-1.5 Seyferts or LINERs) showed behaviour that is similar to that seen in XRBs. An additional two sources (the Seyfert-1.2 NGC 3516 and the LINER-1.9 NGC 3998) showed the expected positive correlation at high $\dot{m}_\mathrm{E}$ and anti-correlation at low $\dot{m}_\mathrm{E}$, but the gradients were considerably steeper than those found in XRBs. The fifth source, the LINER-1.9 galaxy NGC 1052, has a spectrum that is extremely hard and shows no correlation between $\dot{m}_\mathrm{E}$ and $\Gamma$; for this source, as well as for the other unusually hard source NGC 3516, a patchy or variable absorber has been proposed to explain its properties (e.g. Turner et al. 2011; Weaver et al. 1999).

The four AGN sources that are also members of the Palomar survey (namely NGC 1052, NGC 3516, NGC 3998 and NGC 5548) have $L_{2-10\,\mathrm{keV}}/L_{[\mathrm{O\,III}]}$ ratios that are between 56 and 1,300, which in Chap. 2 I found to be generally consistent with emission from Compton-thin AGN. However, despite this positive suggestion of AGN activity, each of the five sources that have been examined have properties (such as spectral hardness and the gradient of the $\dot{m}_\mathrm{E}/\Gamma$ correlation) that are quite distinct from the others. The difficulties encountered when attempting to perform this kind of analysis on AGN extend beyond the low count rates and patchy absorbers, and sources such as NGC 2992 and NGC 4051 were omitted from this study because of uncertainty over the correct model with which to fit the soft excess found in these sources. Despite these problems, NGC 5548 and NGC 7213 show $\dot{m}_\mathrm{E}/\Gamma$ correlations that are remarkably similar to those found in the three XRB sources.

## 5.2 Future Work

The CCF of count rate versus $\Gamma$ provides an unexplored and potentially useful tool for investigating hard-state spectral variability, and further work could perhaps help constrain both the geometry of the accretion flow and the nature and origin of the Comptonising electrons. The large number of high time-resolution *RXTE* observations available in the archive means that any research that makes use of this technique is especially suited to the study of XRBs, and much work can be carried out without the need to request further observations. In particular, I have in this thesis identified 67 observations of GX 339-4—95409-01-01-00 (2010 January 12) to 95409-01-14-00 (2010 April 09)—which deserve further study. These data were discussed in Chap. 4 (see Figs. 4.5 and 4.6) and are unusual in that they exhibit a very tight $\dot{m}_\mathrm{E}$ –$\Gamma$ correlation without the need to add spectra together to improve the signal-to-noise ratio. Overall, the observations extend over an order of magnitude in accretion rate and trace almost the entire range of $\dot{m}_{\rm E}$ between the $\dot{m}_{\rm crit}$ boundary and the (fairly sudden) transition from the hard to the intermediate state. Although no accretion disc is present in the 3–20 keV energy range of these data, the positive $\dot{m}_{\rm E}$–$\Gamma$ correlation implies the presence of a disc that may, or may not, be experiencing changes in the truncation radius. The typical hard-state behaviour and positive $\dot{m}_{\rm E}$–$\Gamma$ correlation exhibited by GX 339-4 during this time is not found in Cygnus X-1, which instead begins to make the transition to the soft state at an accretion rate $\dot{m}_{\rm H \rightarrow HI} \approx \dot{m}_{\rm crit}$, and the increase in the equivalent width of the iron line with accretion rate (see Fig. 4.6; this behaviour has not been observed in my observations of Cygnus X-1 or in those of GX 339-4 by Nowak et al. 2002) further increases the importance of examining these data.

In hardness-luminosities diagrams many XRBs exhibit a characteristic "q"-shaped curve (the hysteresis curve; see GX 339-4 in Fig. 4.10), caused by the different paths taken on this plot when entering and leaving the soft state, such that the hard- to hard-intermediate-state transition in GX 339-4 ($\dot{m}_{\rm H \rightarrow HI}$) occurs at a significantly higher accretion rate than the reverse hard-intermediate- to hard-state transition ($\dot{m}_{\rm HI \rightarrow H}$). If the 67 observations of GX 339-4 discussed above represent the period just before a hard- to hard-intermediate-state transition, then it would be interesting to compare the evolution of the CCFs from these data with those produced from data taken as the source exited the soft state. During these times, we would expect the accretion flow to take the form of a two-zone accretion model with both RIAF and thin-disc components present, and the CCFs to be dominated by the positive correlation. However, Kylafis et al. (2012) argued that the familiar two-zone accretion model with an inner RIAF and outer thin-disc may become reversed, such that the geometrically thick flow appears first on the outer edge of the accretion flow. If this remarkable claim is true then we may perhaps expect to see the positive component of the CCFs become narrower as it weakens, instead of broader as we might otherwise expect.

Although the opportunity for further XRB research using the techniques discussed in this thesis is encouraging, the potential for further investigation into AGN provides less cause for optimism. Of the five sources that were included in Chap. 4, none has more than ~1,400 observations in the *RXTE* archive, which, although sufficient for producing plots of $\dot{m}_{\rm E}$ versus $\Gamma$, is too few to enable an analysis of the count rate versus $\Gamma$ CCFs. Furthermore, the problem of variable absorption in AGN and its effect on measurements of $\Gamma$ suggests that a reliable analysis of AGN spectral variability cannot be performed without including the soft X-ray band (0.3–3 keV) in order to obtain an accurate measurement of $N_{\rm H}$ from each spectrum. Accurately fitting absorption additionally requires long exposure times in order to avoid excessive binning and maintain maximum spectral resolution. These restrictions preclude the use of the *RXTE* archive, which is only sensitive above ~3 keV and tends to provide only short 2–3 ks snapshots when conducting long-term monitoring campaigns, and obtaining a sufficient number of observations from *Chandra* or *XMM* would be impossible given the high demand for these instruments' time. Therefore, any attempt to apply the count rate versus $\Gamma$ CCF technique to AGN will probably have to wait until a new X-ray timing instrument is developed.

Instead, any continuation of the AGN research in this thesis would likely involve the acquisition of further optical/X-ray data in order to provide more information on the nature of LLAGN. A comparison of the absorption corrected X-ray luminosity with various optical line strengths (i.e. the $L_{X}$ v $L_{[O\,III]}$ plot) seems to have much potential in breaking the Compton-thick/starburst degeneracy amongst low-luminosity sources, but to improve significantly upon the results from Chap. 2 (and references therein) would require *simultaneous* X-ray and optical observations in order to reduce the scatter in the plot.

## References

Bohlin, R. C., Savage, B. D., & Drake, J. F. (1978). A survey of interstellar H I from L-alpha absorption measurements. II. *ApJ*, *224*, 132–142.

Kylafis, N. D., Contopoulos, I., Kazanas, D., & Christodoulou, D. M. (2012). Formation and destruction of jets in X-ray binaries. *ApJ*, *538*, A5.

Nowak, M. A., Wilms, J., & Dove, J. B. (2002). Coronal-temporal correlations in GX 339–4: hysteresis, possible reflection changes and implications for advection-dominated accretion flows. *MNRAS*, *332*, 856–878.

Panessa, F., & Bassani, L. (2002). Unabsorbed seyfert 2 galaxies. *A&A*, *394*, 435–442.

Turner, T. J., Miller, L., Kraemer, S. B., & Reeves, J. N. (2011). X-ray characteristics of NGC 3516: A view through the complex absorber. *ApJ*, *733*, 48.

Weaver, K. A., Wilson, A. S., Henkel, C., & Braatz, J. A. (1999). X-ray emission from the prototypical LINER galaxy NGC 1052. *ApJ*, *520*, 130–136.

# Appendix A
# Spectra of Compton-Thick AGN

This appendix contains the 0.3–10 keV *Chandra* images and X-ray spectra of 11 sources that show at least one of the chosen indicators of Compton thickness, namely a flat spectrum above 2 keV and a large iron K$\alpha$ line equivalent width. Only sources with at least 100 counts in the 0.3–10 keV range are shown.

In many of the observations the nuclear region contains numerous point sources. To identify which of these sources coincides with the central black hole, several different approaches have been taken. Firstly, data from the *Hubble* WFPC2 instrument were downloaded (where available) and the position of the optical nucleus was compared to the positions of the X-ray point sources. If this approach was unsuccessful (which was usually the case in edge-on spirals) then the literature was searched for previous studies that may have identified the nuclear emission in some other way (such as radio maps), and these references are provided next to the appropriate plots (Fig. A.1).



C. Skipper, *Fast Spectral Variability in the X-ray Emission of Accreting Black Holes*, Springer Theses, DOI 10.1007/978-3-319-09587-5

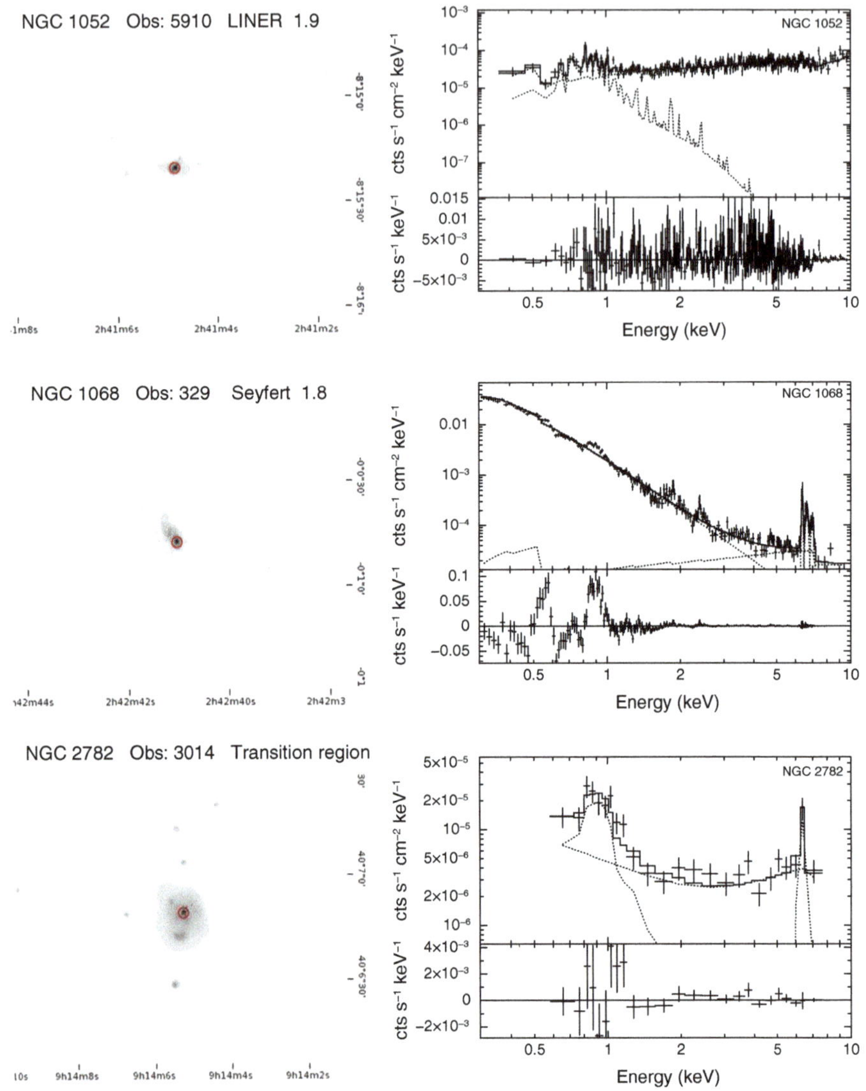


**Fig. A.1** *Chandra* images and X-ray spectra (0.3–10 keV) for the 11 sources that show at least one of the chosen indicators of Compton thickness (which are a flat spectrum above 2 keV and a large iron K$\alpha$ line equivalent width). Only sources with at least 100 counts in the 0.3–10 keV energy range are shown. *Chandra* images and X-ray spectra of 11 Compton-thick sources. *Chandra* images and X-ray spectra of 11 Compton-thick sources. NGC 1068 and NGC 7479 were modelled with an absorbed power law (ZPOWERLW), a reflection component (PEXRAV) and one or more Gaussian components (ZGAUSS) representing the iron lines. NGC 2782 and NGC 4111 were modelled with PEXRAV + MEKAL + ZGAUSS. NGC 3079, NGC 3690 and NGC 4102 were modelled with ZPOWERLW + ZGAUSS. NGC 5194 was modelled with ZPOWERLW + MEKAL + ZGAUSS. NGC 1052 and NGC 4036 were modelled with PEXRAV + MEKAL. NGC 4501 was modelled with PEXRAV

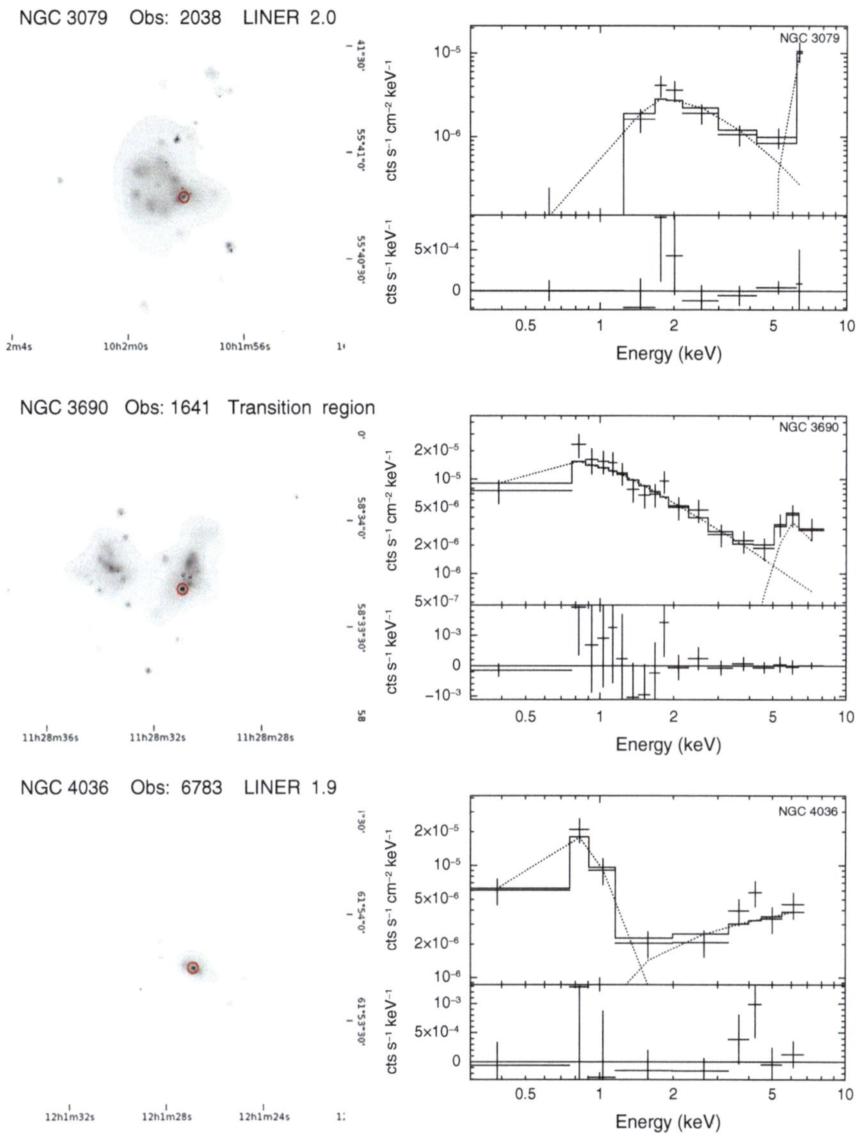


**Fig. A.1** (continued)

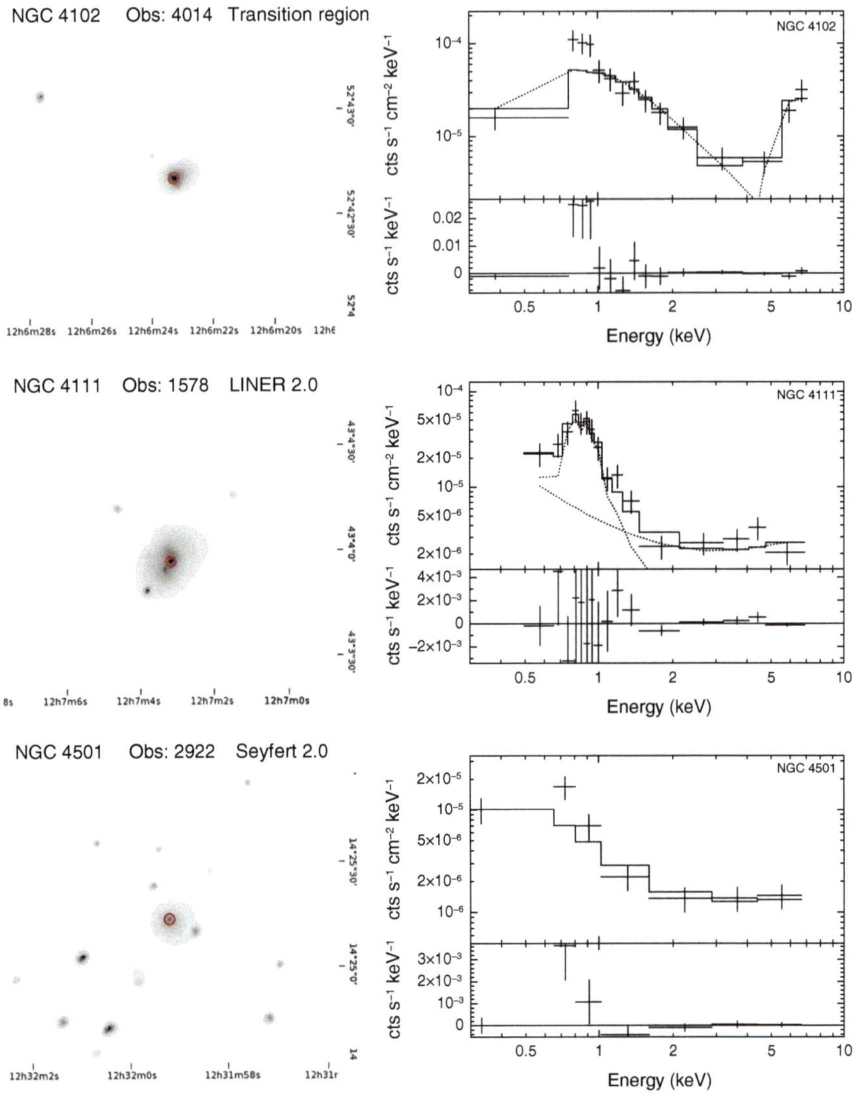


**Fig. A.1** (continued)

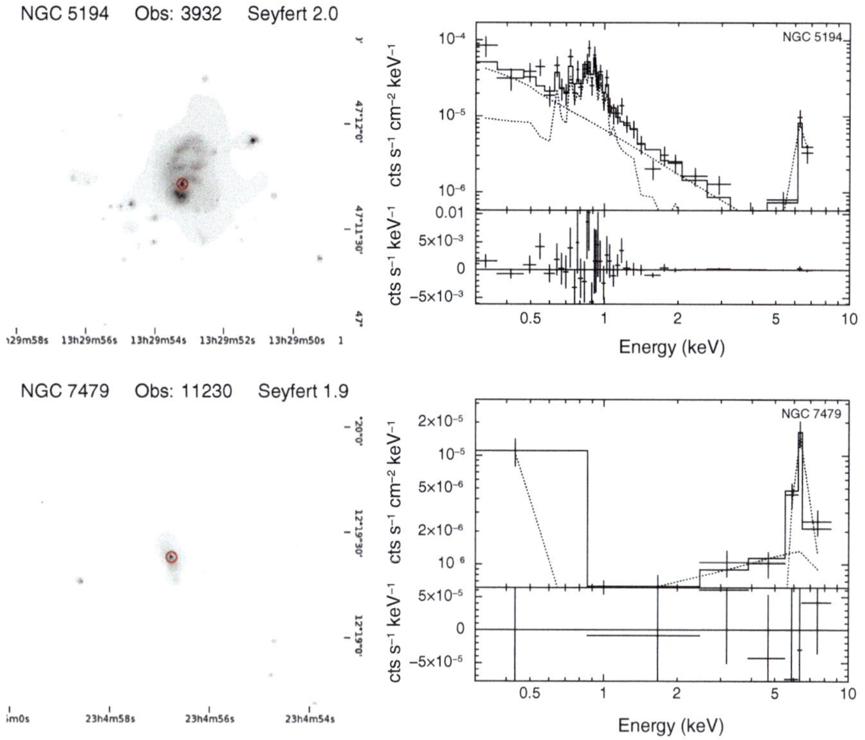


**Fig. A.1** (continued)

# Appendix B
# Spectra of Compton-Thin AGN

This appendix contains the 0.3–10 keV *Chandra* images and X-ray spectra of 56 sources that lack both of the chosen indicators of Compton thickness, namely a flat spectrum above 2 keV and a large iron K$\alpha$ line equivalent width. Only sources with at least 100 counts in the 0.3–10 keV range are shown (Figs. B.1).

Some sources, such as NGC 1275, NGC 3031, NGC 3227, NGC 3516, NGC 4051, NGC 4579, NGC 5033 and NGC 5548 show evidence of pileup in their spectra. This is usually characterised by a turn-up in the continuum above ~7 keV and has been fitted in these spectra using the XSPEC component PILEUP.



C. Skipper, *Fast Spectral Variability in the X-ray Emission of Accreting Black Holes*, Springer Theses, DOI 10.1007/978-3-319-09587-5

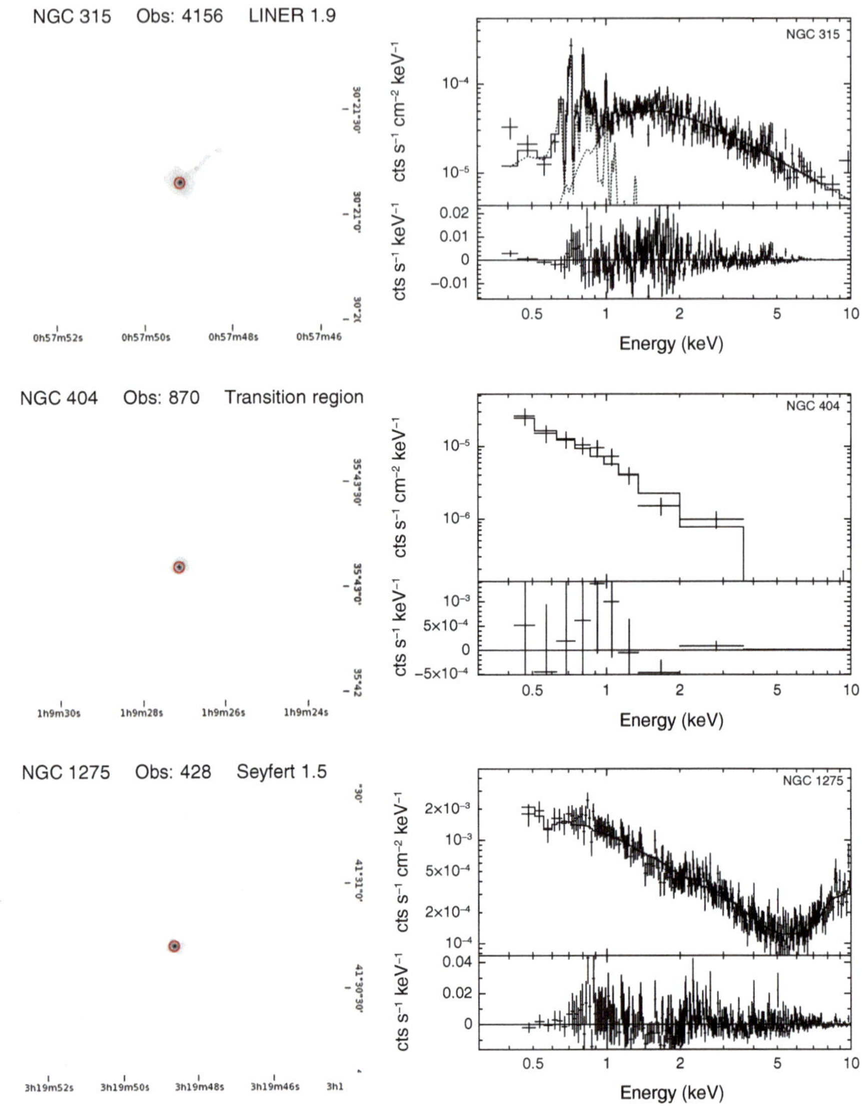


**Fig. B.1** *Chandra* images and *Chandra* X-ray spectra (0.3–10 keV) for the 56 sources that lack both of the chosen indicators of Compton thickness (which are a flat spectrum above 2 keV and a large iron K$\alpha$ line equivalent width). These spectra have been fitted with an absorbed power-law (ZPOWERLW) and, if required, an additional thermal component (MEKAL) to model the soft emission. Only sources with at least 100 counts in the 0.3–10 keV energy range are shown. *Chandra* images and X-ray spectra of 56 Compton-thin sources

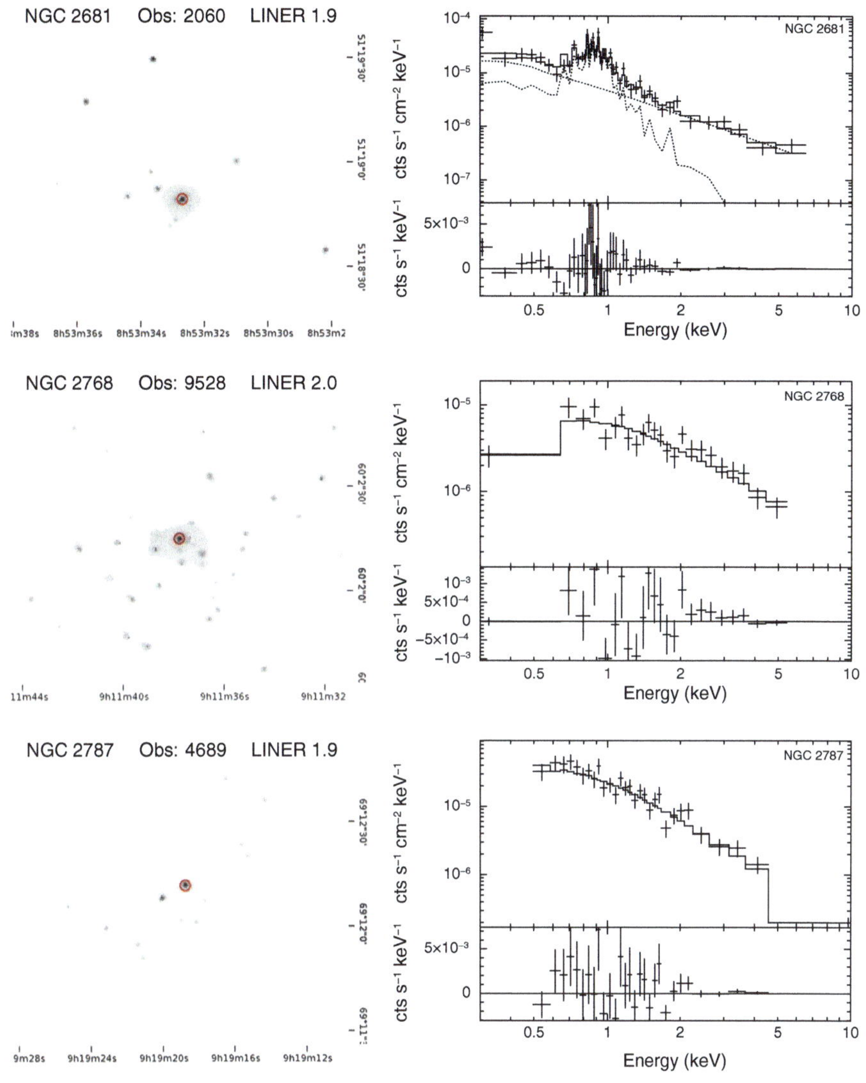


**Fig. B.1** (continued)

NGC 2841 Obs: 6096 LINER 2.0

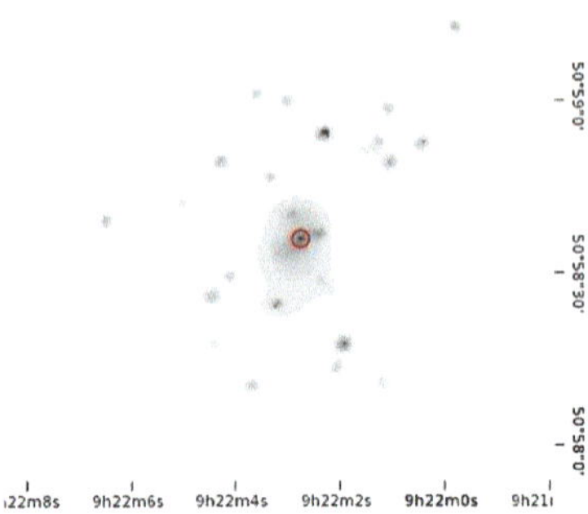


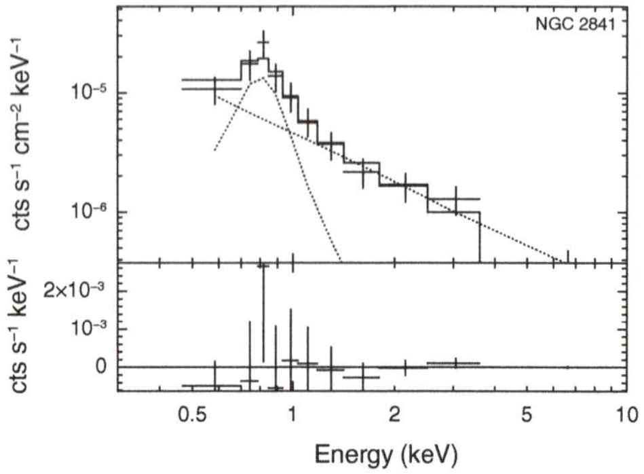


NGC 3031 Obs: 5601 LINER 1.5

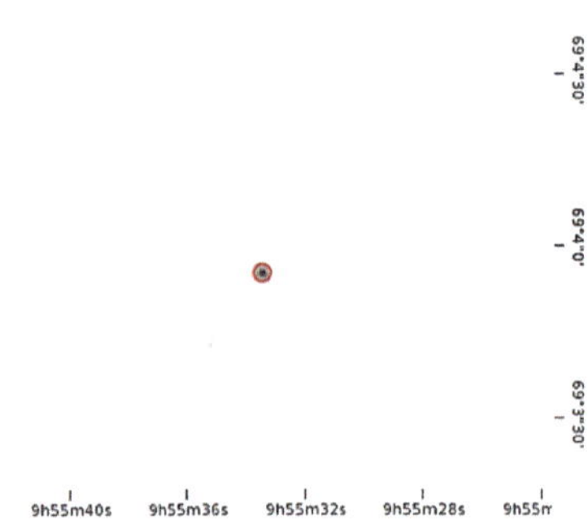


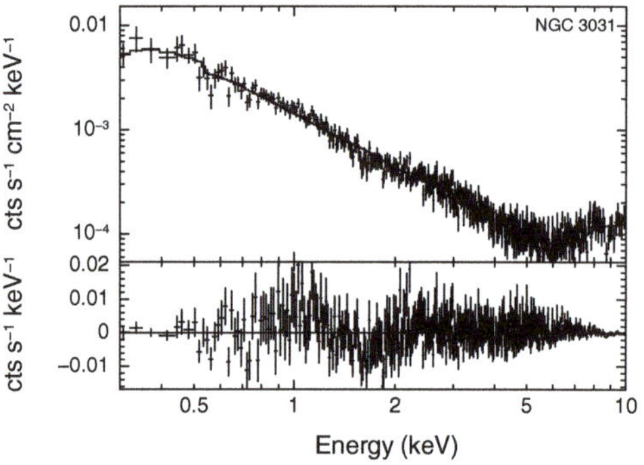


NGC 3077 Obs: 2076 H II region

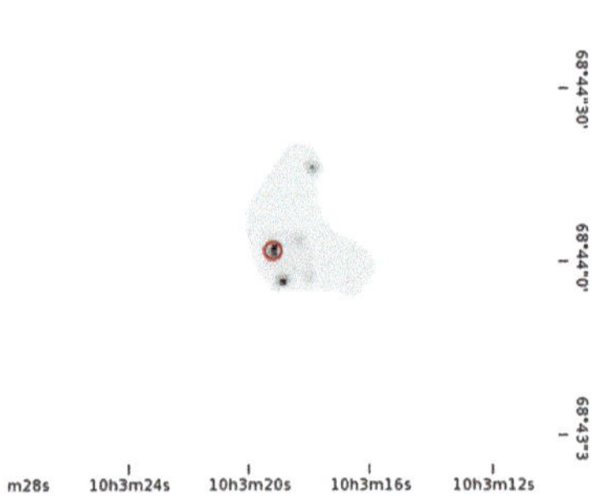


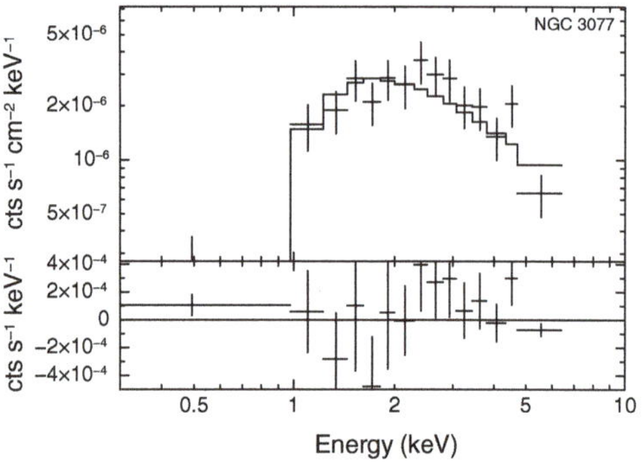


**Fig. B.1** (continued)

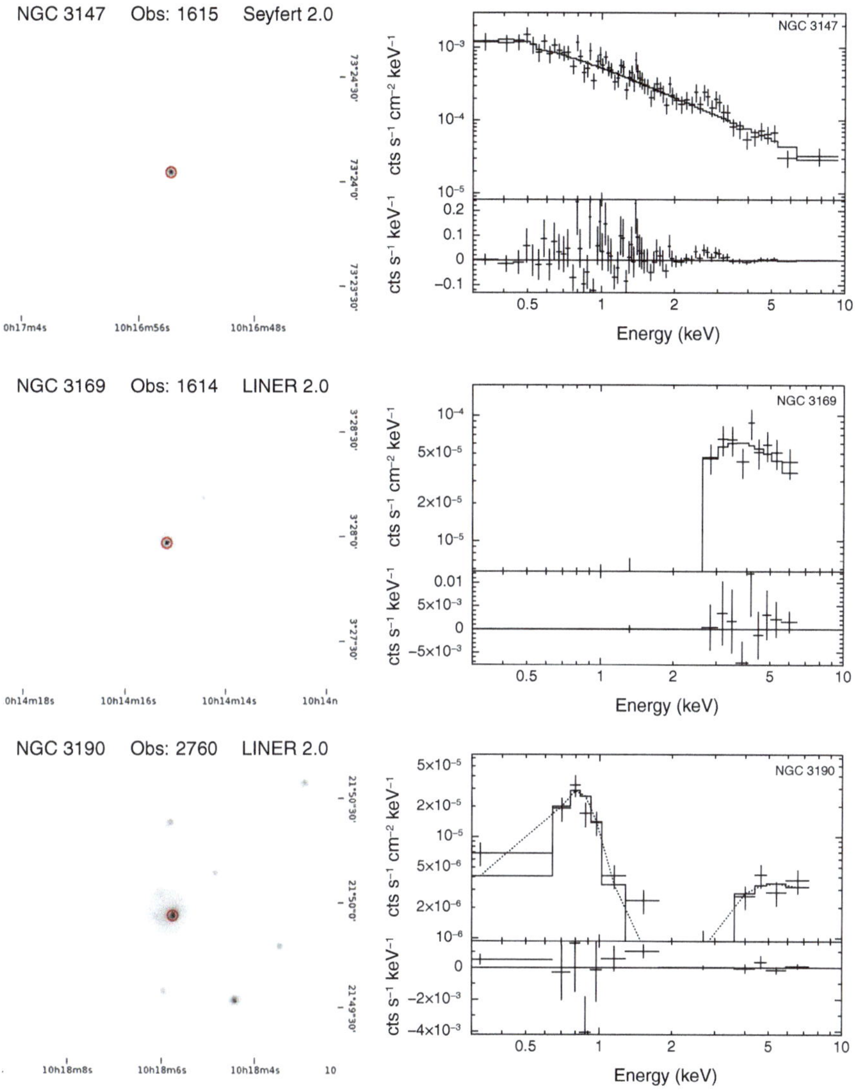


**Fig. B.1** (continued)

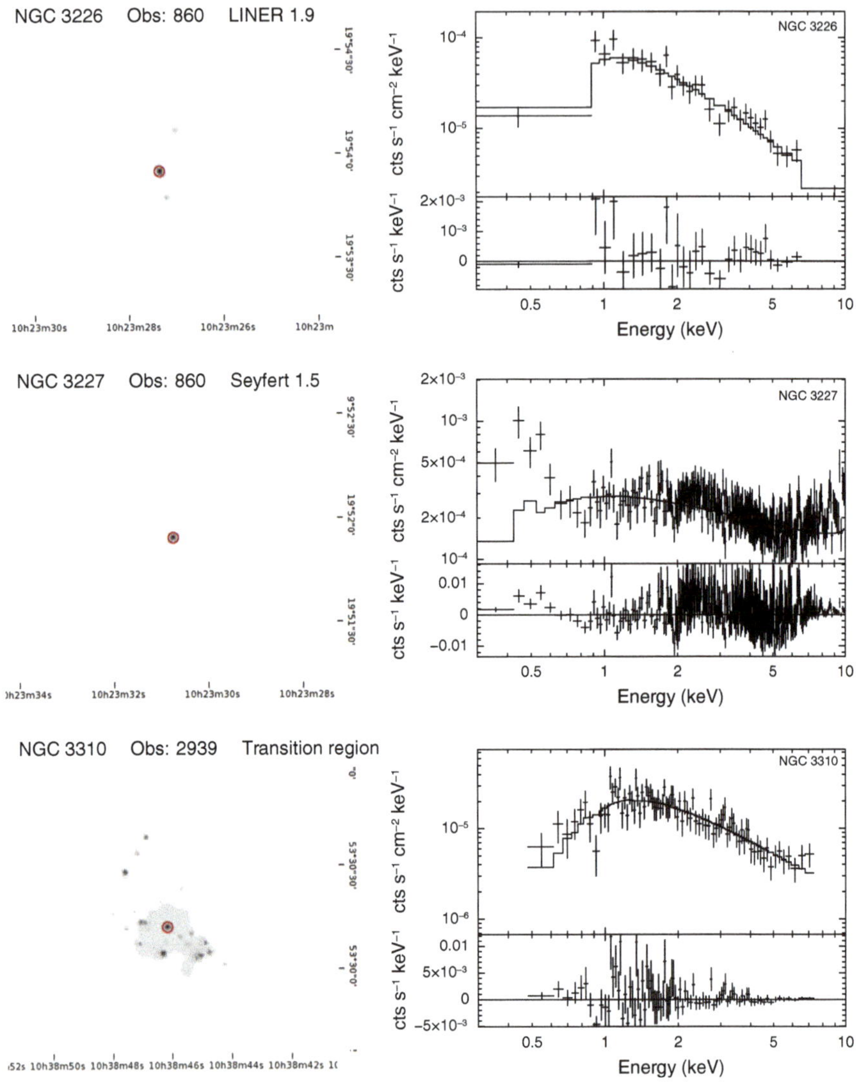


**Fig. B.1** (continued)

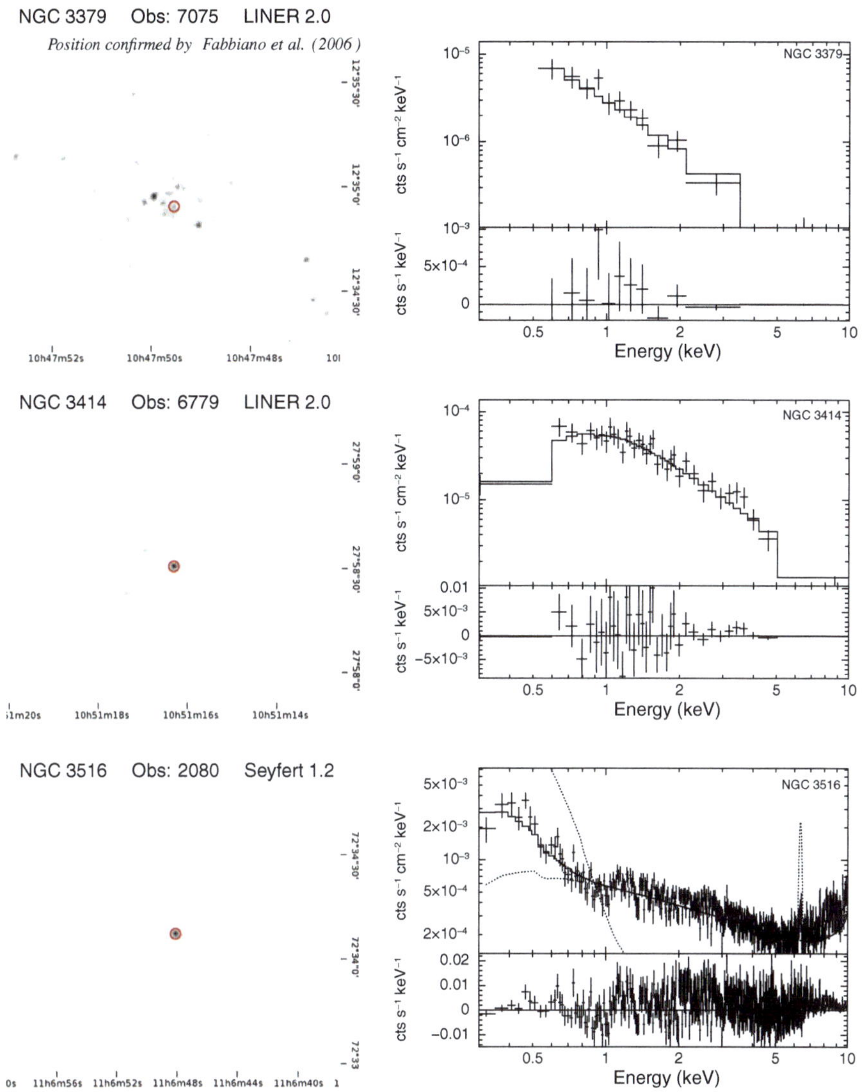


**Fig. B.1** (continued)

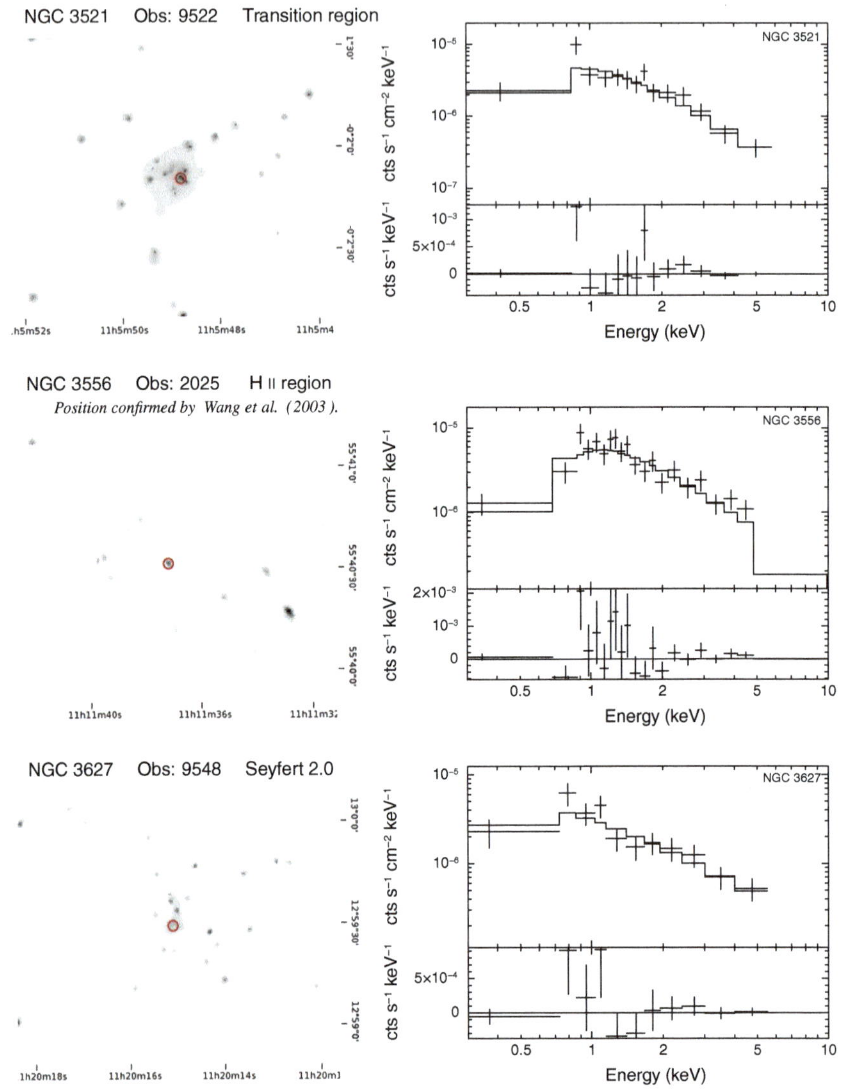


**Fig. B.1** (continued)

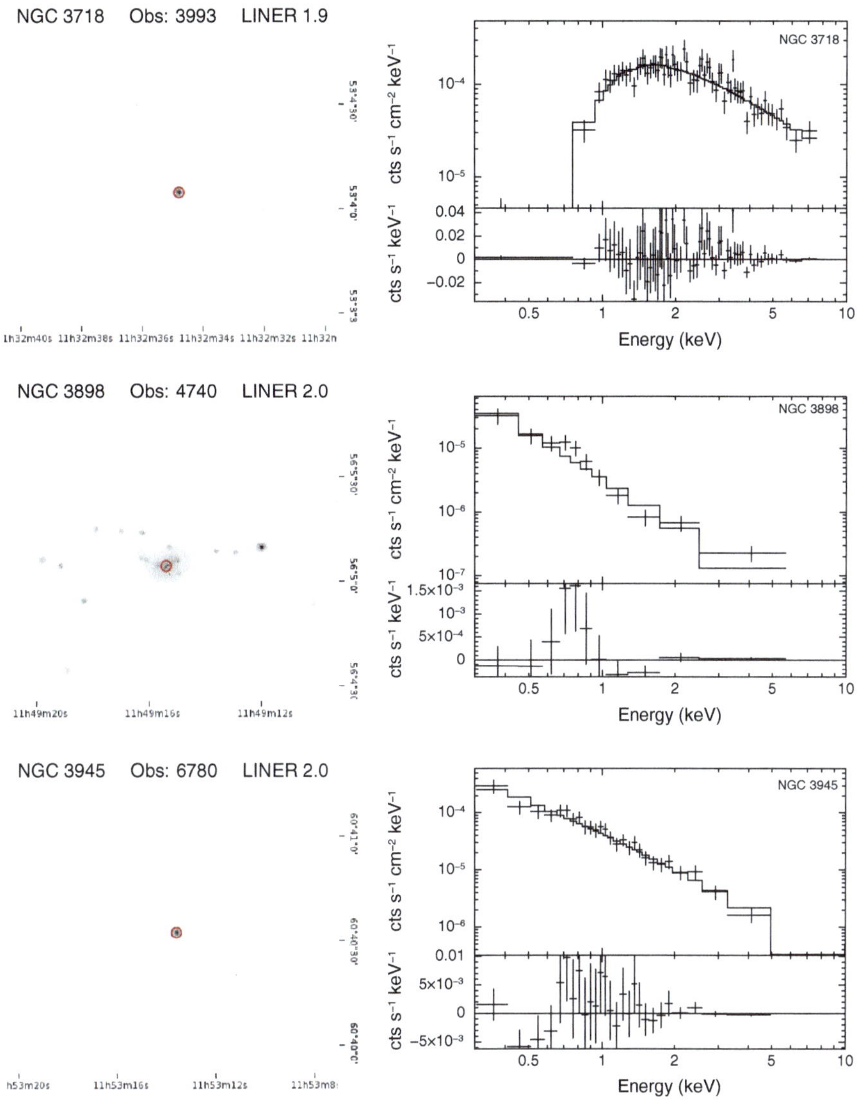


**Fig. B.1** (continued)

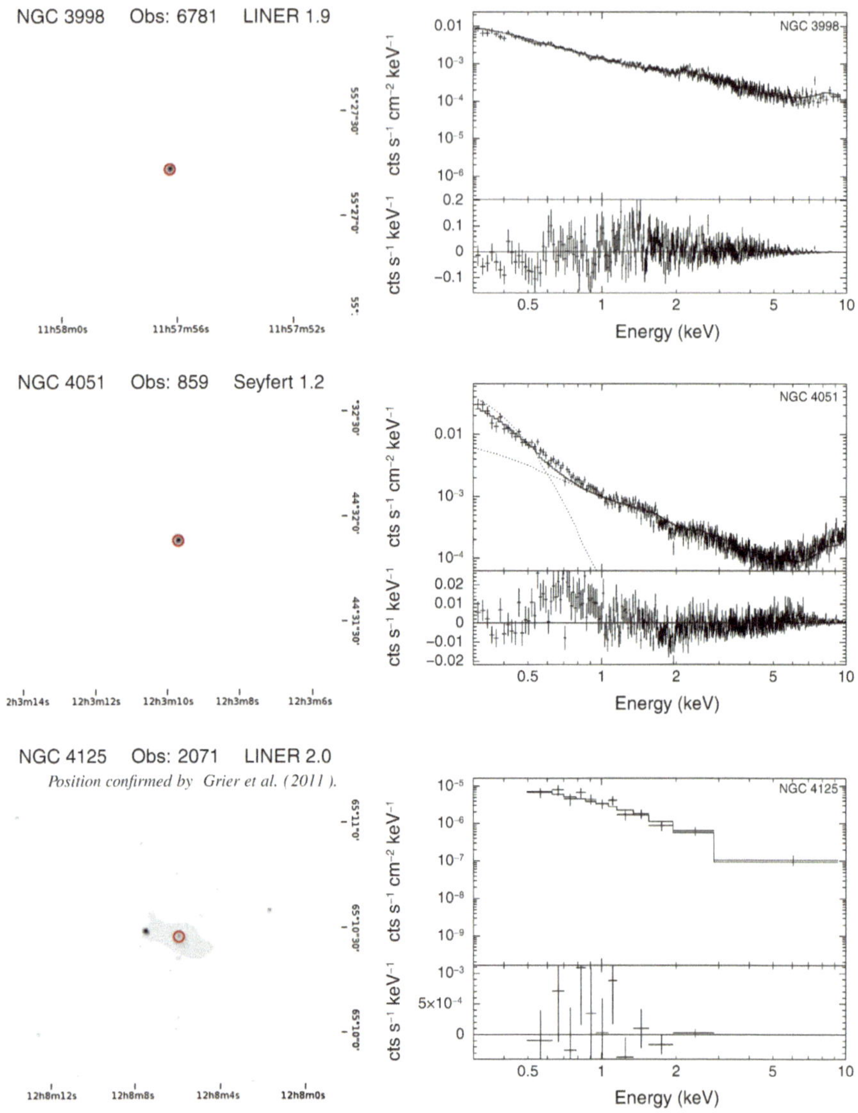


**Fig. B.1** (continued)

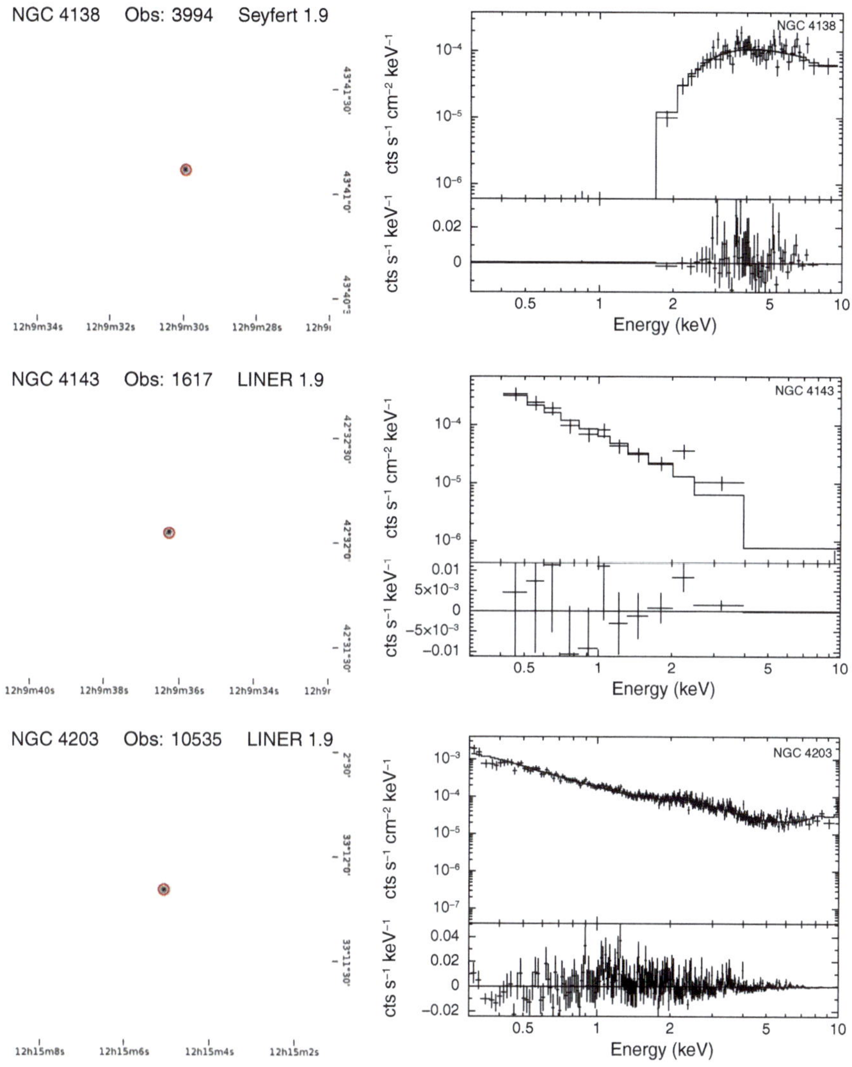


**Fig. B.1** (continued)

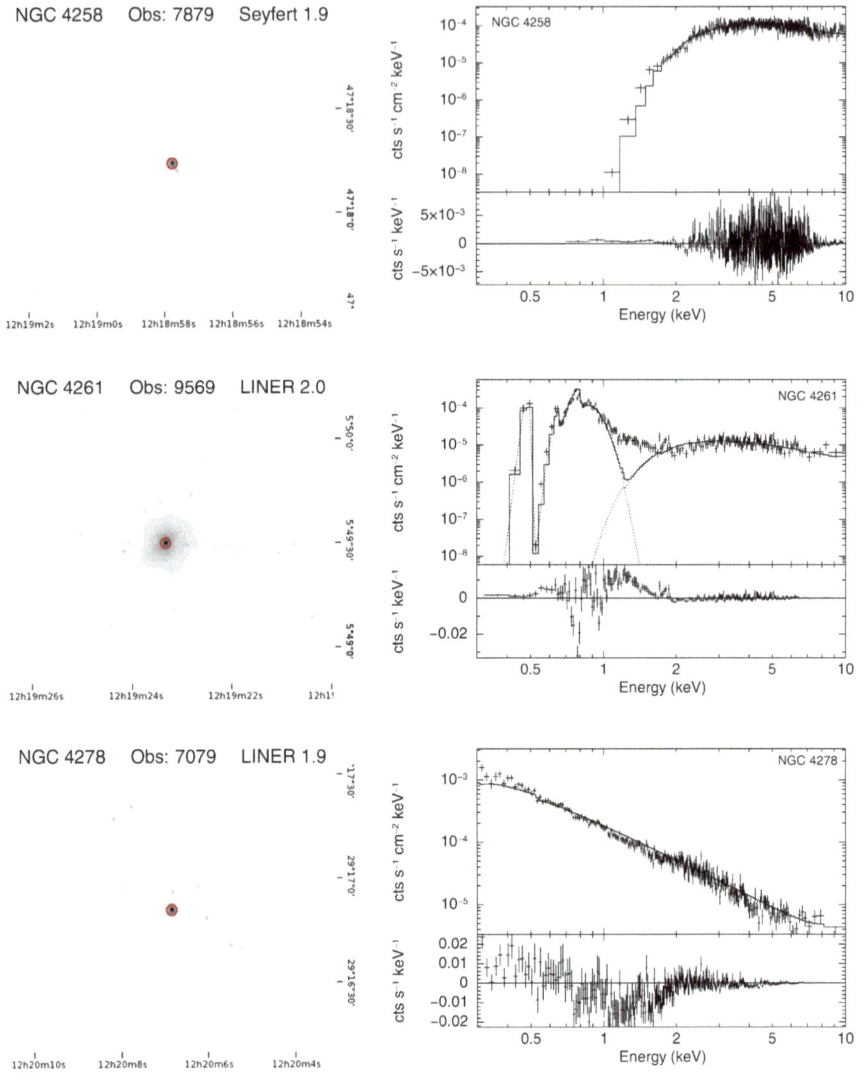


**Fig. B.1** (continued)

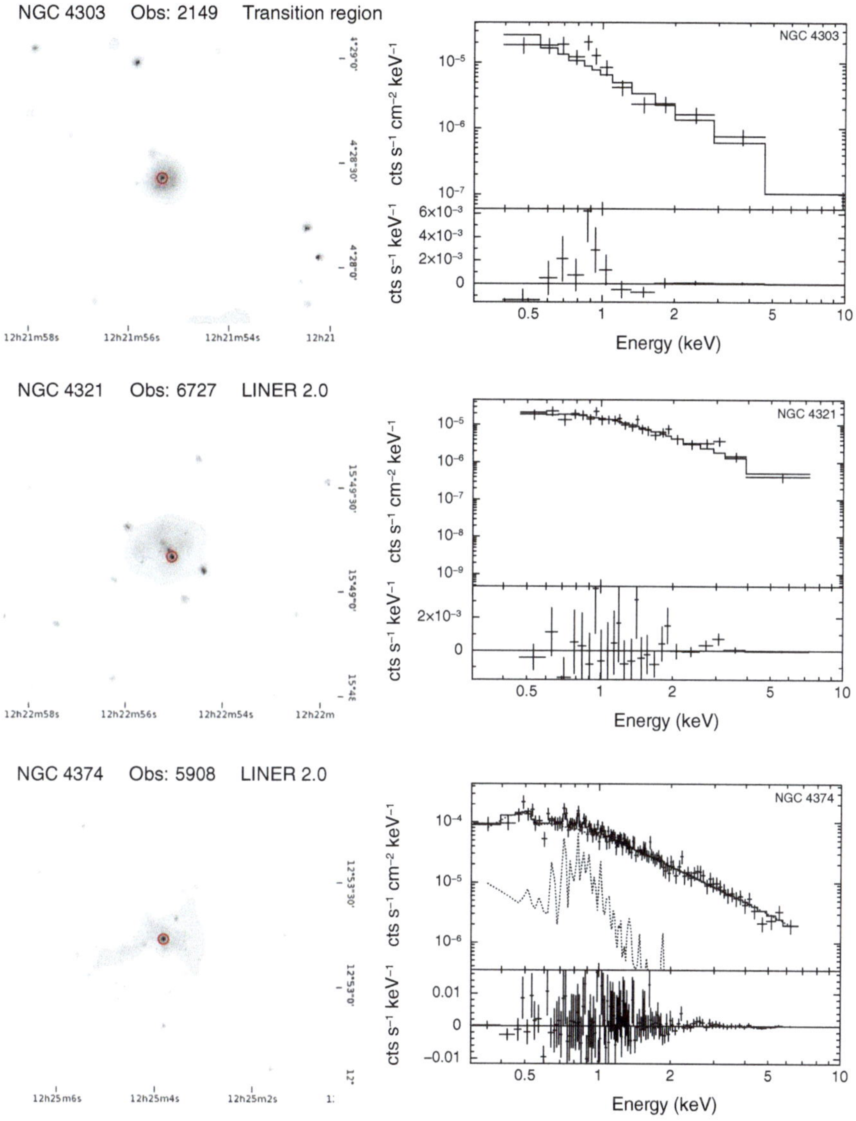


**Fig. B.1** (continued)

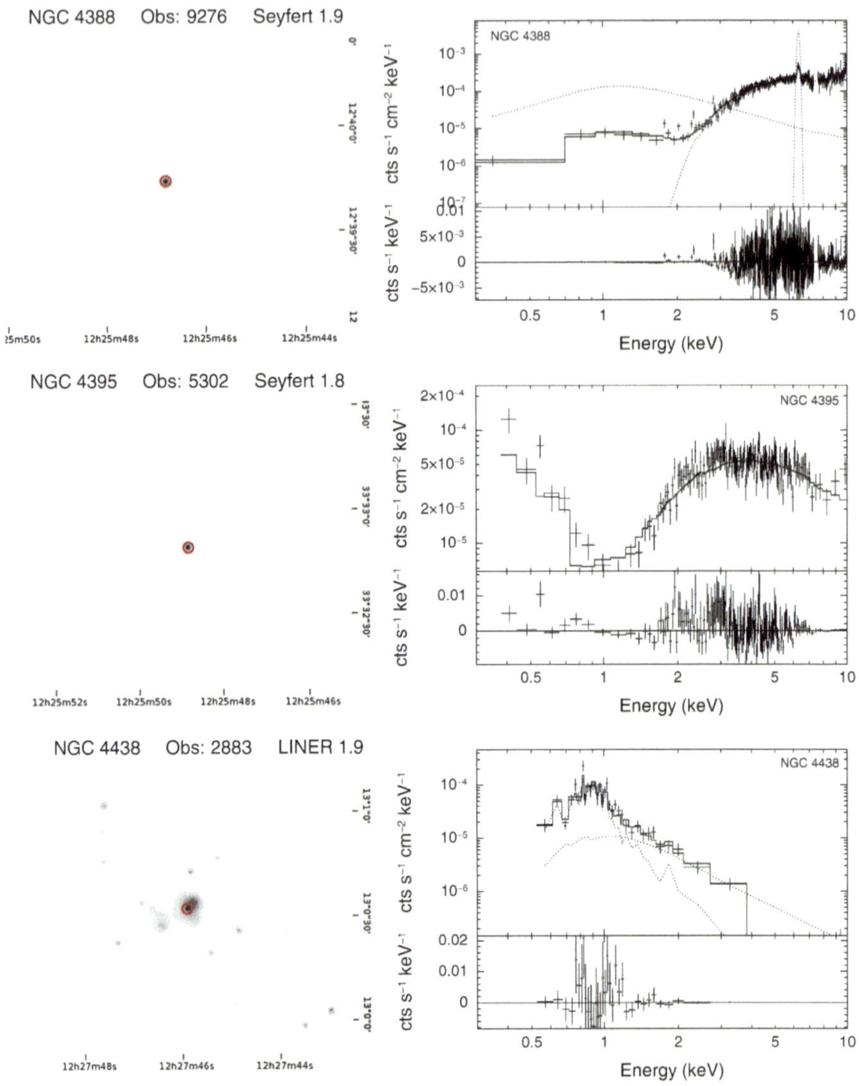


**Fig. B.1** (continued)

NGC 4450 Obs: 3997 LINER 1.9

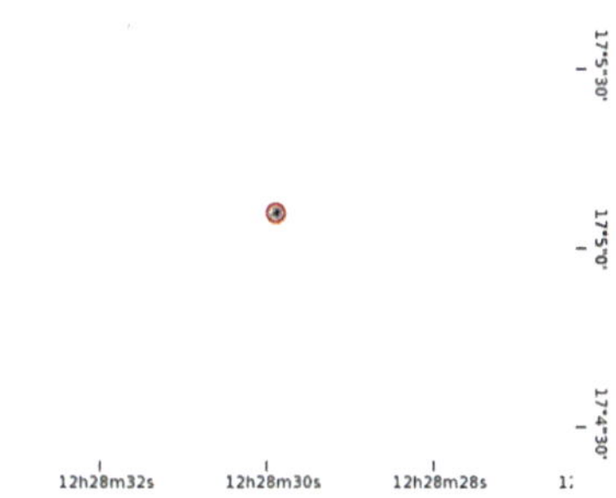


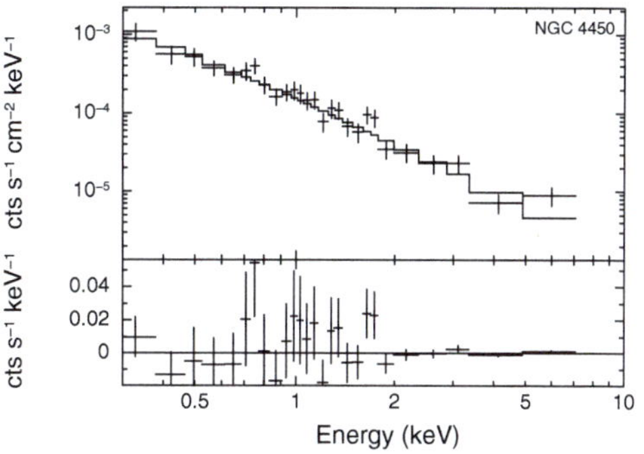


NGC 4457 Obs: 3150 LINER 2.0

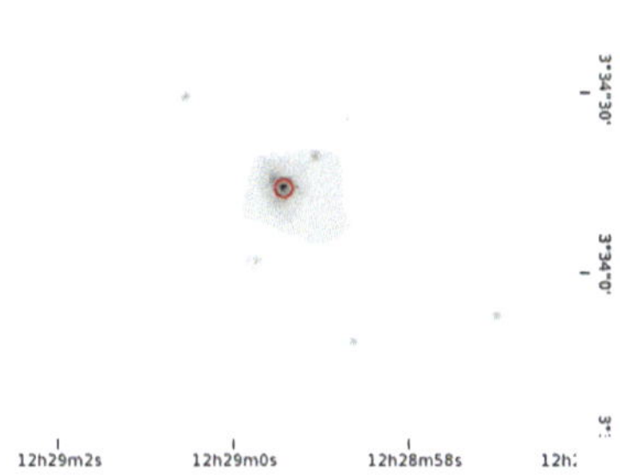


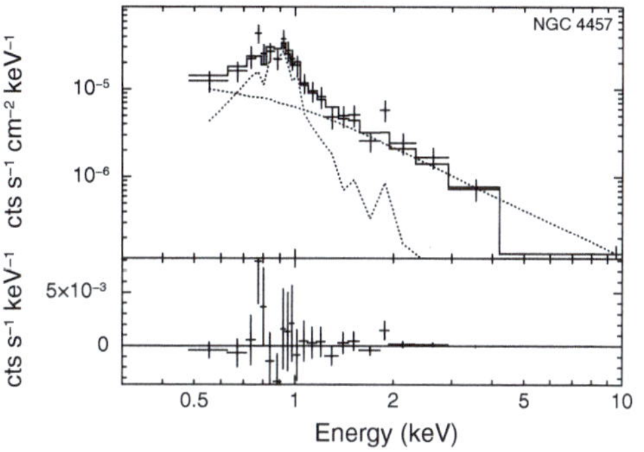


NGC 4477 Obs: 9527 Seyfert 2.0

*Position confirmed by Cappi et al. (2006).*

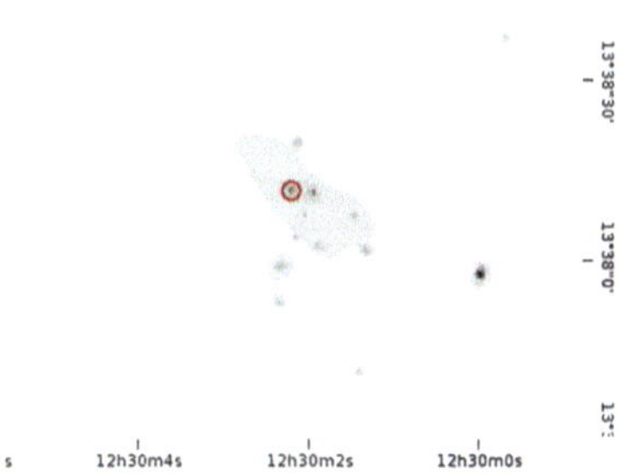


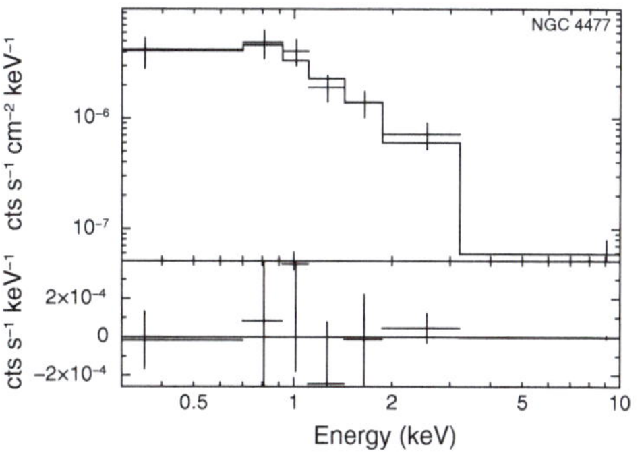


**Fig. B.1** (continued)

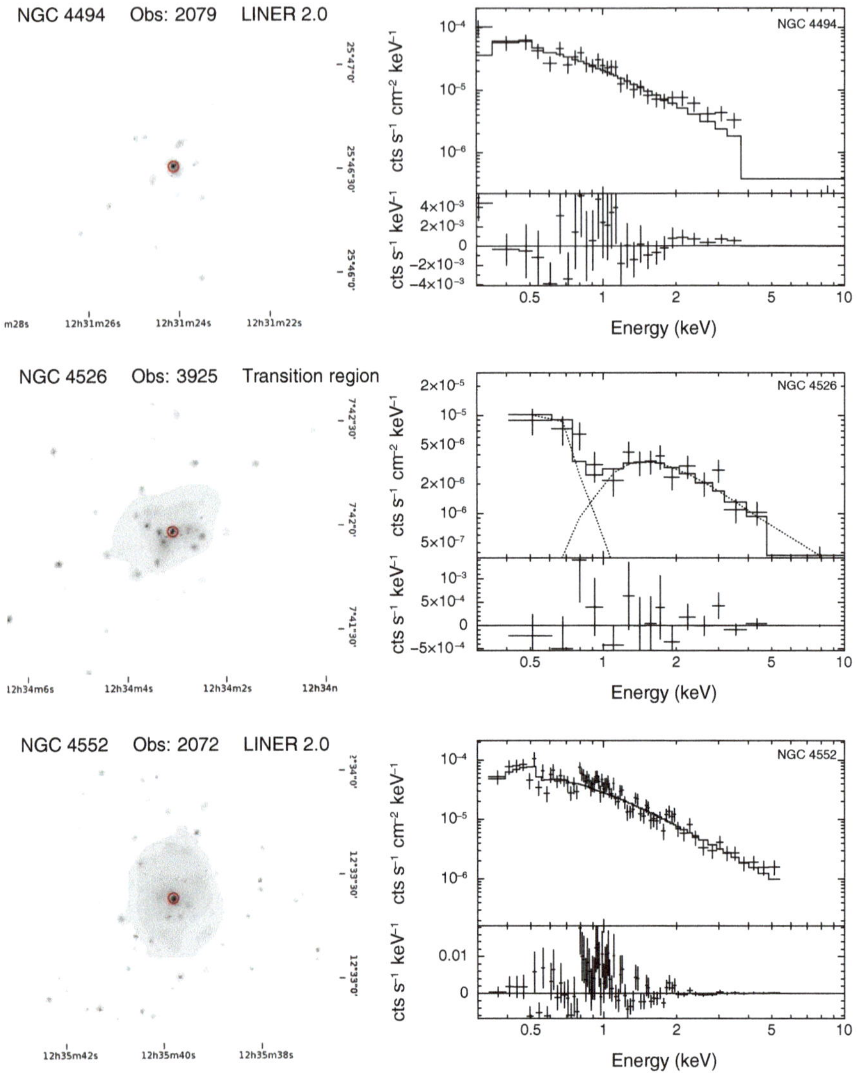


**Fig. B.1** (continued)

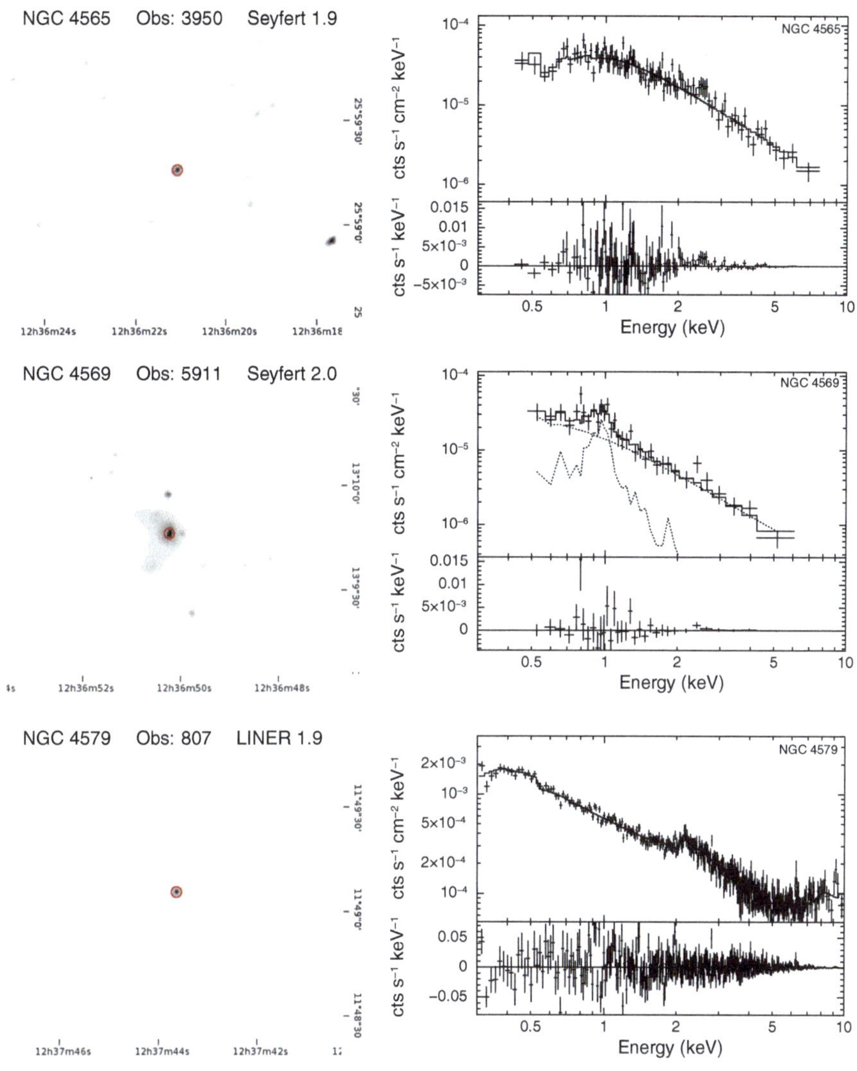


**Fig. B.1** (continued)

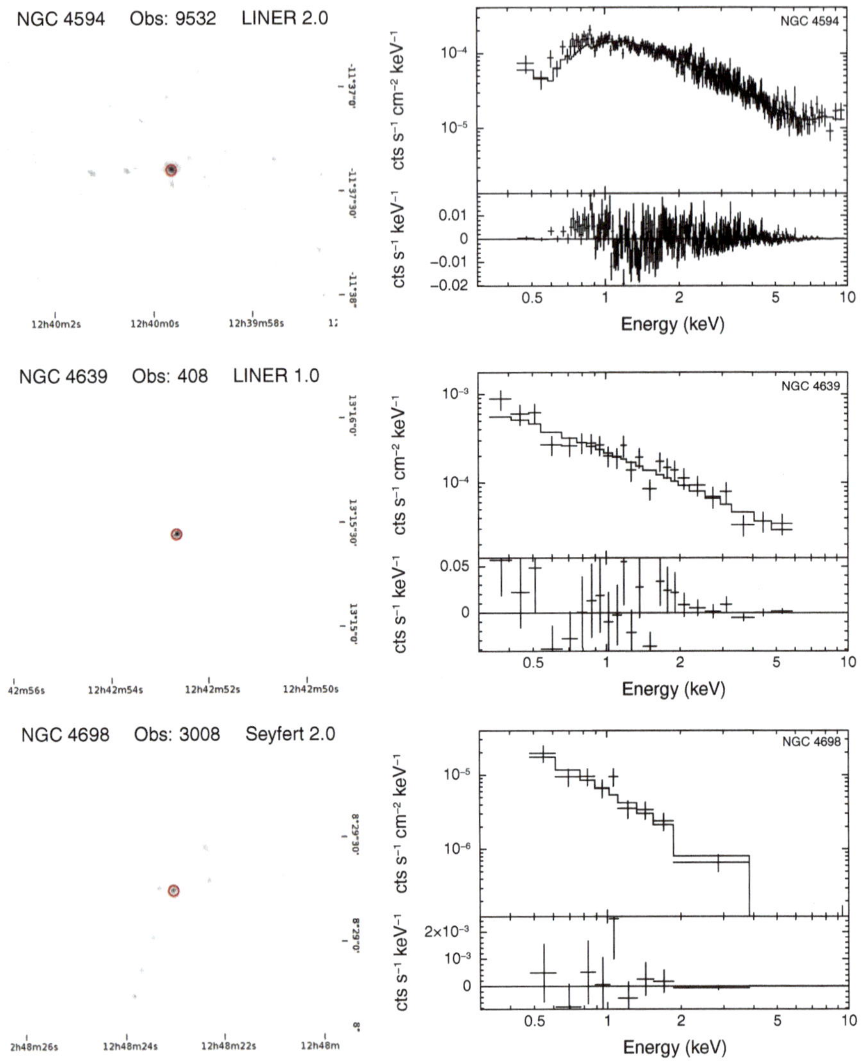


**Fig. B.1** (continued)

NGC 4736 Obs: 808 LINER 2.0

*Position confirmed by Pellegrini et al. (2002) and Eracleous et al. (2002).*

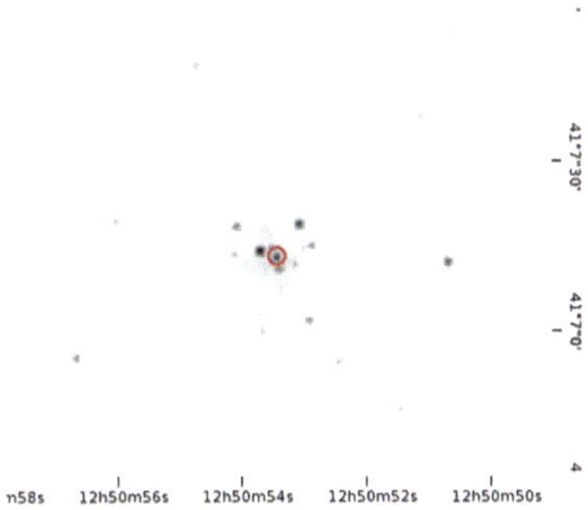


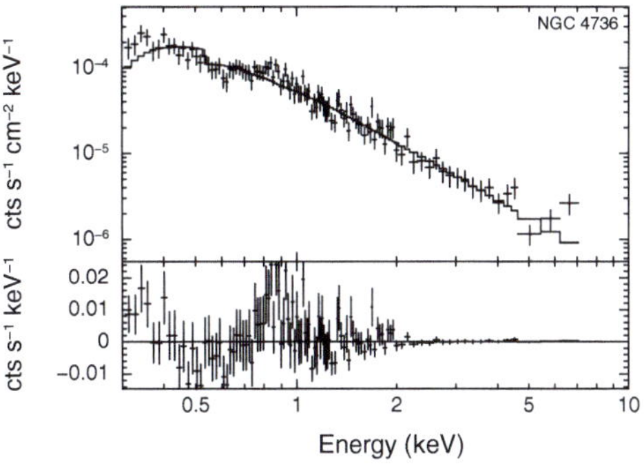


NGC 4750 Obs: 4020 LINER 1.9

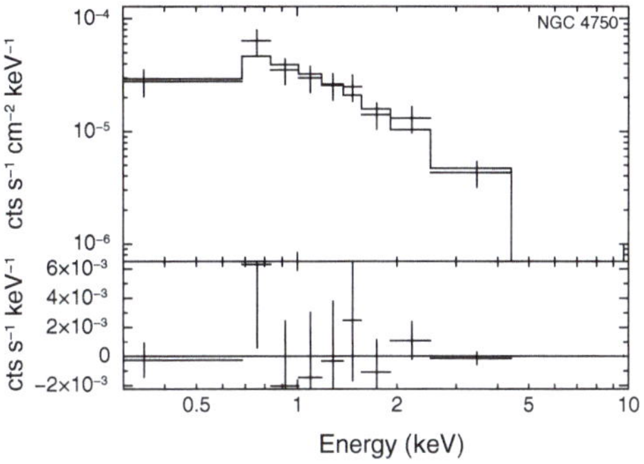


NGC 4772 Obs: 3999 LINER 1.9

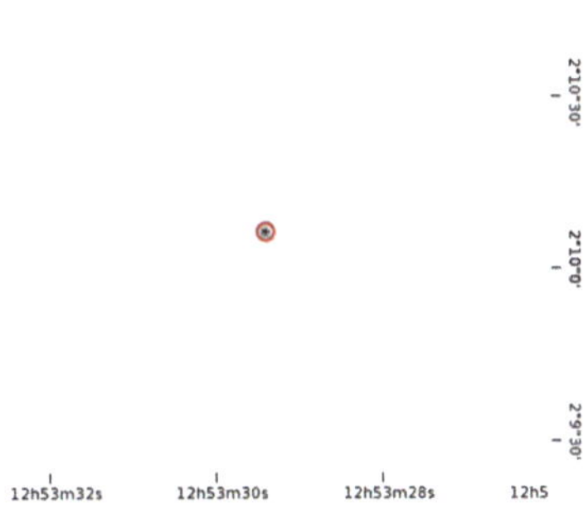


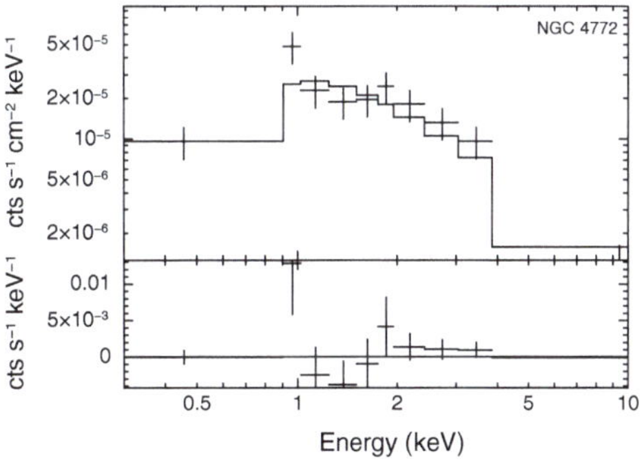


**Fig. B.1** (continued)

NGC 5746 Obs: 3929 LINER 2.0

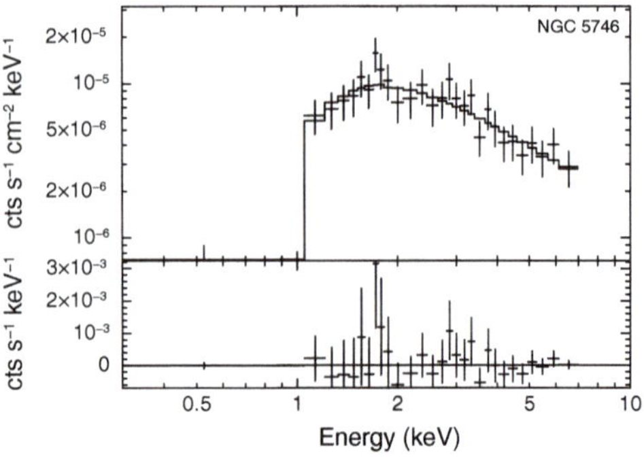


NGC 5879 Obs: 2241 LINER 2.0

*Position confirmed by Zhang et al. (2009). Γ fixed to 2 due to low photon counts and high absorption.*

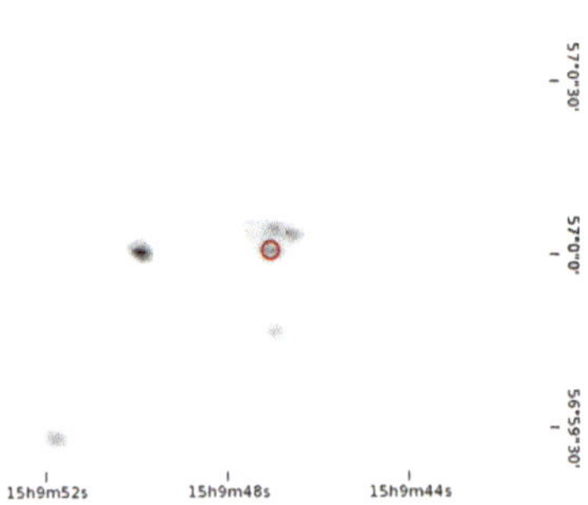


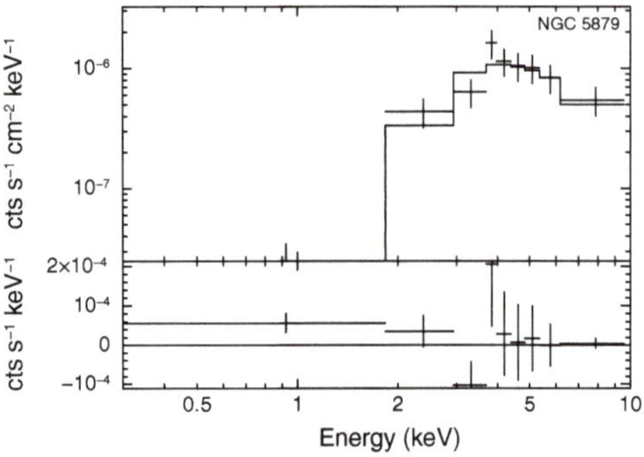


NGC 6946 Obs: 1043 Transition region

*Position suggested by Holt et al. (2003), but not with complete certainty.*

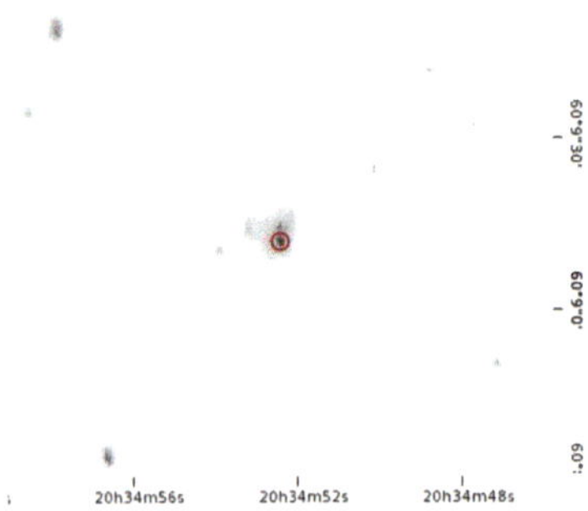


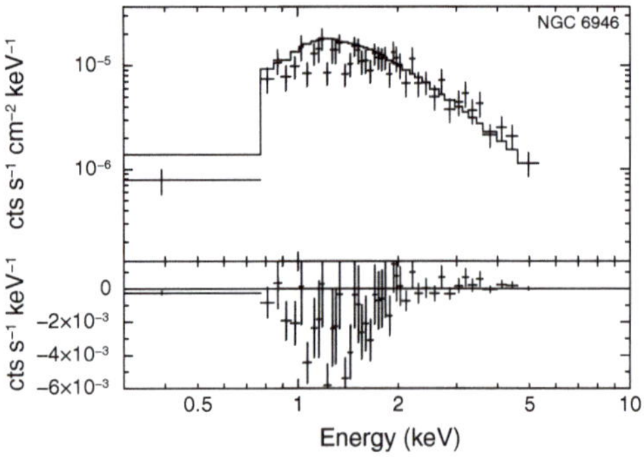


**Fig. B.1** (continued)

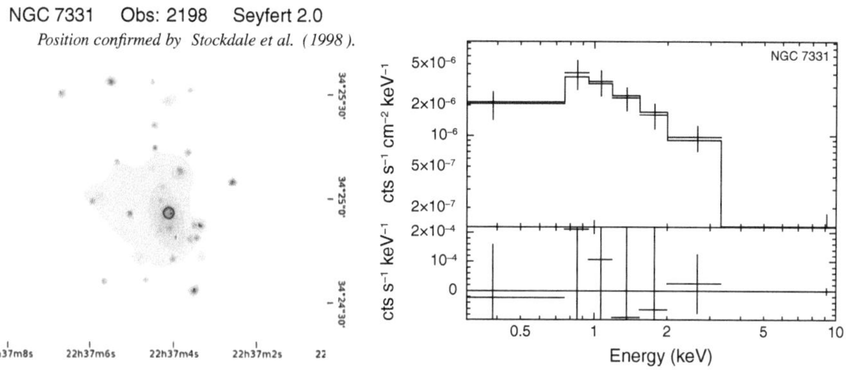


**Fig. B.1** (continued)

# References

Cappi, M., Panessa, F., Bassani, L., Dadina, M., Di Cocco, G., Comastri, A., et al. (2006). X-ray spectral survey with XMM-Newton of a complete sample of nearby Seyfert galaxies. *A&A*, *446*, 459–470.

Eracleous, M., Shields, J. C., Chartas, G., & Moran, E. C. (2002). Three LINERs under the Chandra X-Ray microscope. *ApJ*, *565*, 108–124.

Fabbiano, G., Kim, D.-W., Fragos, T., Kalogera, V., King, A. R., Angelini, L., et al. (2006). The modulated emission of the ultraluminous X-Ray source in NGC 3379. *ApJ*, *650*, 879–884.

Grier, C. J., Mathur, S., Ghosh, H., & Ferrarese, L. (2011). Discovery of nuclear X-ray sources in sings galaxies. *ApJ*, *731*, 60.

Holt, S. S., Schlegel, E. M., Hwang, U., & Petre, R. (2003). Chandra observation of the X-Ray source population of NGC 6946. *ApJ*, *588*, 792–800.

Pellegrini, S., Fabbiano, G., Fiore, F., Trinchieri, G., & Antonelli, A. (2002). Nuclear and global X-ray properties of LINER galaxies: Chandra and BeppoSAX results for Sombrero and NGC 4736. *A&A*, *383*, 1–13.

Stockdale, C. J., Romanishin, W., & Cowan, J. J. (1998). Discovery of a nuclear X-Ray source in NGC 7331: Evidence for a massive black hole. *ApJL*, *508*, L33–L36.

Wang, Q. D., Chaves, T., & Irwin, J. A. (2003). Chandra observation of the edge-on galaxy NGC 3556 (M108): Violent galactic disk-halo interaction revealed. *ApJ*, *598*, 969–981.

Zhang, W. M., Soria, R., Zhang, S. N., Swartz, D. A., & Liu, J. F. (2009). A census of X-ray nuclear activity in nearby galaxies. *ApJ*, *699*, 281–297.

# Index





C. Skipper, *Fast Spectral Variability in the X-ray Emission of Accreting Black Holes*, Springer Theses, DOI 10.1007/978-3-319-09587-5

**L**



**N**



**O**



**P**



**Q**



**R**



**S**



**T**



**X**